THE ALL-ROAD BIKE REVOLUTION

MAKE YOUR BIKE FAST, COMFORTABLE AND RELIABLE

JAN HEINE

With a foreword by Ted King

Illustrations by MIYOSHI

Bicycle Quarterly Press

Contents

Part 3

Sidebars

Foreword

by Ted King

The bicycle is an amazing machine. Its history has run from the draisienne, or two-wheeled hobby horse, to the velocipede and the penny farthing, and to dozens of other iterations of this two-wheeled contraption. And yet the general design that we picture when we hear the word 'bicycle' has not changed in 130 years: two wheels of equal size, connected by a diamond frame and powered by pedals via a chain drive. It remains unchanged because it's well documented as the most efficient machine for traveling in terms of energy consumed for moving a certain distance as a function of body weight.

The bicycle can be collected. It can be beautiful artwork. Getting out on a bike can be a hobby or a source of recreation. It can be a profession in any number of ways, as it was for me in the realm of bicycle racing. The bicycle is a source of community and togetherness, acting as a magnet to bring people together.

Jan Heine, the author of this book, brings people together from all corners of the globe as the editor of Bicycle Quarterly. He and I have shared rides exploring the majestic Cascade Mountains, and we've hosted group rides for all-comers. We've shared countless conversations into seemingly every aspect of the bicycle. We talk about tires, tire pressure, tire width, tire material. We talk about frame material, frame geometry, frame manufacturing. We talk about the physics of cycling from cornering forces to gyroscopic forces. We talk about the real-world applicability of all of these things.

In these hours upon hours of conversation, I've come to realize Jan has one of the most inquisitive and scientific minds towards the bicycle. In these fewer than 250 pages, this book delves into, quite frankly, every facet of cycling. Jan's mind comes to life over the course of this book. The proven science of cycling, backed by fascinating, anecdotal applicability, will be appreciated by everyone. From the first-time bicycle owner to a lifelong cyclist looking to purchase their 20th bike, this book covers it all.

While the wheel has been around for more than 5,000 years, the historic timeline of the bicycle is just a blink of the eye. Despite all the interest to advance technologically, as we have the tendency to do as humans, the bicycle remains a remarkably simple machine. In my two decades in the sport of cycling, I've witnessed many – certainly not all – aspects of cycling, from road bikes to mountain bikes to cyclocross and gravel bikes (all of which are covered in this book, of course). As time goes on, the changes become more and more nuanced. The more things change, the more they stay the same. This book will continue to do what I consider to be the bicycle's greatest ability, and that's to draw people together as they enjoy the amazing thing that is the bicycle.

Prologue

One Bike?

Is one bike all we need? Can the all-road bike replace the racing bike, the mountain bike and the touring bike that many cyclists used to have in their stables? In fact, it already has: These days, the Bicycle Quarterly team uses the same bikes for a fast-paced Saturday morning ride through the hills that surround Lake Washington; for a 36-hour adventure exploring the highest passes of the Cascade Mountains; and even for an overnight camping trip to the end of the gravel road at Drunken Charlie Lake. Fast pavement, rough gravel, even with a camping load – all-road bikes are our preferred choice because their performance is more accessible and comes with fewer compromises.

The incredible versatility of modern all-road bikes has not just changed our bikes, but also the way we ride. We now enjoy the spirited performance that we used to associate only with racing bikes – all the time. We are no longer limited to smooth pavement as we search for that great feeling of gliding along effortlessly. This has freed us to explore new routes away from traffic and crowds, and it's allowed us to experience new ways of cycling. This book is about the science that has made this 'All-Road Bike Revolution' possible. It's about getting more enjoyment from cycling. It's about bikes, but it's really about creating wonderful experiences.

Comfort or Performance?

This question used to guide our choice of bicycle. A racing bike with narrow tires offered great performance, but it wasn't very comfortable. A touring bike gave you a comfortable ride, but at the expense of speed. A mountain bike had wide tires to go off-road, but it was ill-suited for spirited rides on pavement.

All that has changed: We now know that comfort and performance are not opposed, but inextricably linked. Vibration costs energy. A smoother ride is faster. Tour de France racers now ride on tires as wide as the touring tires of the past. All-road bikes combine the speed of racing bikes with the comfort of touring bikes and the wide tires of mountain bikes.

Choosing your bike is no longer about compromise. Modern all-road bikes are comfortable *and* speedy. Their wide, supple tires roll equally fast on smooth *and* rough surfaces. All-road bikes transcend the traditional categories. Now we use the same bike for spirited road riding, for gravel, even for touring!

More than ever, choosing your bike and its parts carefully is essential, so it can fulfill the promise of comfort *and* speed. This book will help you understand why some bikes perform better than others, and how to find one that performs best for you. For your current bike, you'll read how to improve its comfort, performance and reliability.

For almost two decades, Bicycle Quarterly's research has been at the forefront of the all-road revolution. Our research starts and ends with riding bikes: how to make our bikes faster, more comfortable, more reliable – more fun. This book combines everything we've learned into a coherent, easily digested format. It contributes many new insights about bicycle design and performance. It confirms some long-held beliefs. And it relegates others to the realm of myths once and for all.

Some of what you will read here goes against what is still accepted as common knowledge, yet it is based on careful research performed to high scientific standards. It's also been tested on the road. Just fifteen years ago, our findings that wider tires roll as fast as narrow ones were met with disbelief. The idea that stiffer frames don't always perform better also took some time to gain widespread acceptance. So did new ideas about bicycle geometry. Our understanding of how bicycles work has evolved in recent years. We'll continue to refine this knowledge in the future. What you will find in this book is based on the latest research. It forms the basis for a exciting new breed of all-road bicycles that no longer require you to choose between comfort and performance.

Why Performance?

Why do we care about the performance of our bikes? Why do we spend many hours with research, testing and fine-tuning? Why do we obsess over the smallest details when we spec our bikes? Does it even matter?

We think it matters. Cycling elevates humans into the realm of birds. We fly along on the flats. We soar up the mountains as if in the updraft of thunderclouds. We swoop downhill like a raptor in a steep dive. We feel the wind in our faces, we tuck for descents, and we spread our wings on uphills. In a group, we play around one another like a flock of starlings. The more performance our bikes offer, the more we get to experience these joys.

Performance does not necessarily mean absolute speed, competition or racing. Performance is about feeling our bodies become one with the bike – not about dropping the next rider. Performance puts a smile on the face of a child when he or she first rides away, unshackled from the steadying hand of a parent. Performance is what makes cycling fun.

It is this spirited riding that this book is about. There are many reasons to cycle. Most of us ride because it's fun. Some ride purely for transportation. Others do it for health reasons, or to lose weight. All riders benefit from better performance. The more you enjoy your bike, the more you will ride it!

In the past, we thought that choosing performance meant giving up comfort and versatility. We now know this isn't true: Comfort *is* performance. This means performance is available to everybody, not just those willing to suffer in exchange for a little extra speed. This realization has revolutionized bicycle design to the benefit of all cyclists.

A third important component of bicycle design is reliability. A bike that breaks down costs time and money in repairs. If you are unlucky, it leaves you stranded far from home. In the worst case, it can cause a crash. Fortunately, performance and reliability both are the result of good design, so they usually go hand in hand.

A final consideration is aesthetic. A beautiful bike not only is nice to look at, it usually performs better, too. The old saying "What looks right is right" applies to bicycles as much as it does to anything else.

These four points outline what is important in a bike: It should be fast, comfortable, reliable and beautiful. Through our research, we've learned that these factors are not independent variables at odds with each other, but parts of a coherent whole.

Ideally, a bike is inexpensive, too, and we'll discuss which compromises don't affect the performance of your bike, and where it's worth spending the money for high-quality equipment. The goal of this book is not to tell you what bike to ride, but to provide you with the information you need to make your own choices.

In Part 1, we start with a brief guide that summarizes what you need to know when choosing a bike. This part is intentionally written to be useful for new cyclists and experienced riders alike. Part 2 gets into more detail as it covers the science of how to make a bike fast, comfortable, reliable and fun. Part 3 discusses how to apply these findings to the individual parts of a performance bike.

In this book, we focus on practical advice that has proven itself on the road, not on theoretical reasoning that may or may not have applications in the real world. Our tests and measurements always start with real-world observations, which we then confirm in carefully controlled settings. Bicycle Quarterly's goal is to get rider and bike working together in complete harmony, because that is what makes spirited riding so much fun.

Part 1

Choosing

Your New Bike

You are reading this book, so I don't need to tell you why you need a bicycle. Yet choosing one can be an overwhelming experience. You quickly learn that a bike's brand doesn't mean much. Most companies offer bikes at every price point and quality level. How can you tell whether a frame is good? What should you look for in components: good tires or a high-end rear derailleur? What about saddle and handlebars? Which parts are easy to change and which will remain with you as long as you own the bike? And which are important?

The following chapter is a quick guide to help you decide which bike is right for you. We start with determining what type of bike you want, then talk about narrowing down the many choices until you can find a bike that will be good for you.

"It is by riding a bicycle that you learn the contours of a country best, since you have to sweat up the hills and coast down them. Thus you remember them as they actually are."

–Ernest Hemingway

Which Type of Bike?

Road bike, mountain bike, city bike, touring bike, gravel bike, adventure bike, cyclocross bike... the choices can be bewildering. Which one is right for you? The bike industry loves to divide bikes into categories, hoping to sell each rider multiple bikes – ideally one in each category.

Asking friends who are experienced cyclists isn't always helpful. Bicycles have been transformed as our understanding of how they work has changed – that is the subject of this book. This also means that much of what we used to 'know' isn't true any longer, yet many cyclists still believe that you have to choose between a fast road bike that you'll ride only for short distances on sunny days or a slow touring bike that can take you on adventures... They'll ask you: "What do you want to do with the bike?"

You'll probably answer: "Everything! I want to take my bike on a nice Sunday morning ride around the city. And maybe I'll also want to join a spirited group chatting and trading pulls as they fly along the road. And sometimes I'll also want to commute by bike. And I've got a dream of going on a multi-day tour." Fortunately, you don't need multiple bikes to do all these things and more.

Just as a modern car can be used for grocery shopping, spirited drives on twisty mountain roads and long road trips, modern all-road bikes excel on almost every ride, from Saturday morning jaunts to continent-spanning adventures. Unless you want a specialized bike for competition or extreme terrain, a modern all-road bike will serve you well. All-road bikes are often marketed as 'gravel bikes,' but they aren't just for riding on unpaved roads.

A good all-road bike combines the performance of a racing bike with the comfort and go-anywhere abilities of wide tires. The result is a bike that's fun to ride on smooth pavement *and* on rough gravel, over short distances *and* really long ones. It's no coincidence that all-road and gravel bikes are the most popular category in the cycling market right now – they bring out the best in bicycles.

Tires

As you begin choosing your bike, your first consideration should be its tires. The tires themselves are easy to change – and you'll probably want to change the inexpensive stock tires of most production bikes – but the frame's clearances determine how wide a tire you can fit. Wider tires roll as fast as narrow ones, yet they offer more comfort and safety. The first question when buying a new bike is: How wide a tire is wide enough?

Unless you ride only the best handmade racing tires on the smoothest roads, don't even consider anything narrower than 32 mm. I ride 42 mm tires on paved roads, and I prefer wider ones for gravel. There is an upper limit: Tires wider than 60 mm are hard to fit between road cranks with a narrow stance (Q factor) to facilitate a smooth spin. For most riders, tires between 38 and 55 mm provide an excellent compromise that will roll smoothly on all types of roads. Wide tires also don't easily fall into cracks in the road and make negotiating railroad tracks much easier. This will make your cycling much safer, too.

Size/Fit

Once you've determined what type of bike you want, you'll start looking at individual models. Now the most important consideration will be the bike's fit. Just like a pair of shoes, your bike must fit you. Otherwise, it will be difficult to realize its potential for performance and comfort. Never buy a bike (or shoes) that don't fit well!

How to determine which bike fits you? You can get a professional fitting. If you already have a bike, you can work with an experienced rider or builder. Have them watch you ride and suggest adjustments until you've found your position.

Fortunately, most of us have relatively average bodies, and bike companies make bikes in sizes that fit 90% of the population – at least for men. Choices for women are improving, too. If you fall outside this range, then a custom bike may be your best choice, rather than buying a bike that is too big or too small and trying to make it fit with an overly short or long seatpost and stem.

Riding Position

Your riding position determines your bike fit as much as the dimensions of your body. Your riding position isn't a matter of style, but it is determined by your power output. Your goal is to put very little weight on your hands, arms and shoulders. When you pedal, your upper body acts as a counterweight to the pedaling forces, which are generated by the legs, and also by your core. The harder you pedal, the more you want to incline your body to remain balanced on the bike.

Without that counterweight, it's difficult to produce much power. That is why all riders bend lower over the handlebars when they pedal harder, and why stronger riders need a lower position than casual riders, and vice versa.

If a rider's position is too inclined for their power output, their arms must support the weight of their torsos. Leaning hard on the handlebars causes discomfort in hands, wrists and shoulders. It also upsets the handling of the bike.

Your core strength plays a role in your riding position, too. The stronger your core, the more inclined your comfortable riding position will be.

Your riding position affects your bike's fit. A more upright position requires a longer head tube and a shorter top tube, to get the handlebars closer to the rider.

Bear in mind that your strength may increase as you ride more, and you may need a lower riding position in the future. Small adjustments can be made by swapping stems for longer and/or lower models, so it might make sense to start out with a relatively short stem. Then you can install a longer stem to adjust the bike as your legs and core get stronger.

Frame

The frame is the heart of the bike. It determines the bike's performance. A good frame moves in sync with the rider's pedal strokes. Try riding several bikes up a moderately steep hill. Perhaps you will notice differences in how the bikes feel and how easy they are to pedal for you. It can take time to get in tune with a bike, so it may be hard to tell which frame is good for you without riding it for a week or two. On a relatively

Pedaling pushes the rider's torso upward. The rider leans forward to provide a counterweight. A casual rider (left) sits more upright, while a stronger rider (right) is more inclined. Both are balanced and comfortable.

affordable bike, a lighter frame tends to perform better. Heavier frames are often too stiff to offer good performance. (It's not the weight, but the right amount of flex, that is important.)

The frame also determines the bike's geometry and thus its handling. Wider tires make the differences between geometries more pronounced, so this topic is becoming more important with modern all-road bikes. Test riding will show you which bikes handle the way you like.

The quality of the frame is harder to determine. Today's manufacturing is so good that blobby welds and gaps in the joints are rare. If a frame fails, it's usually because of invisible flaws in the materials themselves. Fortunately, frame failures are rare, and most makers offer good warranties.

Fork

The fork is as important as the frame. It forms part of the bike's suspension. The tires absorb the vibration of pavement irregularities – often called 'road buzz' – but they cannot absorb bigger bumps, such as broken pavement or steps at the joints between concrete road panels. Good forks can absorb these bigger, less frequent bumps. A test ride will tell you how stiff the fork is. How much do bumps jolt the handlebars? There is no need to suffer on your bike – in Part 2, you'll see why good shock absorption makes your bike faster, too.

Some believe that a stiffer fork results in more precise handling. This is true only when you ride out of the saddle. And even then, the stiffness of the wheels is of greater importance. Ride the bike out of the saddle to see whether the fork flexes too much for your liking – which is very unlikely. When you descend and corner hard, the bike is always balanced (otherwise, you'd fall over), and fork stiffness doesn't affect how the bike handles.

Flat or Drop Handlebars?

We don't always ride at the same speed, so performance bikes have drop handlebars that allow several distinct positions. When going slowly, riders hold the upper parts for a more upright position. To go faster, they use the lower parts of the bars to match their position to the higher power output. Moving your hands between positions also reduces pressure and prevents numbness and injuries.

Flat handlebars are good for shorter rides at constant speeds. City bikes use them. Flat handlebars are light and simple, but they also limit a bike's versatility.

Make sure the handlebar position matches your power output: During your test ride, go at a moderate speed and hold onto the tops of the bars. Try bending your elbows. This should be comfortable. Then accelerate to your cruising speed and put your hands on the brake hoods. Can you bend your elbows without effort? Repeat with a bigger effort and your hands in the drops. If all these positions are comfortable, then the positions match your power output.

Your power will probably increase as you ride your new bike more, so you'll want to start out with a relatively short stem in a high position. Then you can swap it for a longer/lower stem once you develop more strength.

Drop handlebars offer many positions to match the rider's variable power output.

Wheel Size

The gyroscopic force of the wheels stabilizes the bike by centering the handlebars. Smaller wheels and narrower tires make the bike more nimble. Larger wheels and wider tires make it more stable. This allows fine-tuning the handling of a bike, so it's stable enough to be relaxing to ride, yet nimble enough to steer exactly where you want to go.

Many all-road bikes use smaller 650B wheels to compensate for their wide tires. Should you get two wheelsets, one with wide tires for gravel and one with narrow tires for pavement? This isn't necessary: Narrow tires and taller wheels don't roll faster (see Part 2). However, if you want to ride fast racing tires on pavement and sturdier tires on gravel, two wheelsets make it easy to swap between them.

Gearing

Make sure the bike has low enough gears for you. The gearing of many bikes used to be much too high for general riding. Fortunately, that is getting better – many all-road and gravel bikes have low gears that are useful for real-world riding. Unlike in the past, changing the gearing isn't always easy. A bike that doesn't allow you to pedal comfortably up the hills

For the same rider and fit, bicycle frames can have different dimensions, wheel sizes, etc.

will spoil the fun of cycling. Don't assume you need triple cranks to get low-enough gears. Compact and ultra-compact double cranks can offer gears that are just as low, and their shifting performance is often better. 'One-by' gearing with a single front chainring is a good choice if you don't mind larger steps between gears.

Equipment

You may only plan to ride on sunny days, but a bike that is fully equipped with fenders, lights and luggage can take you on adventures beyond your dreams. The small weight penalty of fenders, lights and a bag won't lessen the bike's performance on shorter, faster rides. Even if a ride starts out dry, you might get caught in a thunderstorm and appreciate fenders. Exploring mountain roads sometimes has you run behind schedule and return in the dark. And you'll want to bring spare clothes and maybe even your camera, plus a picnic lunch.

In the past, fully equipped performance bikes simply did not exist. It was ironic: Even the most exotic sports cars are equipped for driving at night and in the rain. Yet the only bicycles with that equipment were city or touring bikes, which didn't offer the spirited performance we are looking for. These bikes were slow, not because they had fenders and a rack, but because their frames weren't optimized for performance.

Some bike shops will tell you that you can retrofit fenders, racks and lights. That is true, but it isn't always as easy as it seems, unless the bike is designed with these components in mind. For fenders, all mounting points should be at the right distance from the tires and allow direct mounting of the fenders without brackets or clamps. A good front rack also requires mounts in the right places. And if you want to install generator lighting, it's best if the bike is prepared for this. Bike makers are beginning to realize this, and performance bikes are starting to be equipped for a wide range of real-world riding.

Less Important: Components

The components of your bike are less important. These days, most derailleurs work well. There are greater differences in brake performance: Check on a test ride that the brakes can stop quickly even when going down steep hills. (Don't forget to shift your weight backward and brace yourself by locking your elbows, so you don't fly forward when braking hard.)

More expensive components usually are lighter (less important) and last longer (more important as you ride more). In general, most frames that offer good performance will be equipped with components that are more than adequate.

Details

Small details can make a big difference in how a bike performs for its rider. A subtle difference in the curve of the handlebars can prevent hand numbness on long rides. A longer front fender can keep the rider's feet and the bike's drivetrain dry and clean in the rain. Details like these are why experienced riders often buy custom bikes, where they can choose every part exactly as they like.

Ordering a custom bike allows you to get every detail exactly the way you want.

Easy to Change: Saddle, Pedals, Tires

Three of the most important components of your bike are easy to change, so they shouldn't influence your choice of bike. Saddles are very personal, and it may take some experimentation to find one that works for you. Most riders have a pedal system they like, and many bikes come with cheap pedals just to let you to test-ride the bike. The assumption is that you'll install the pedals of your choice. Supple high-performance tires make the biggest difference in your bike's feel, comfort and performance. They are a worthwhile upgrade. Budget to replace these parts within the first year of owning the bike.

Budget

How much do you need to spend on your new bike? Like a car, a bicycle is a considerable investment. With cars, even the cheapest ones will work reliably and drive safely, because people who buy them expect to use them day after day. Unfortunately, there are still many bikes that are sold as toys or for occasional use only. Currently, you have to spend about $1,200 to get a bike that will offer you the joys of spirited riding and that will last for more than just a season or two.

Why are even entry-level bicycles so expensive? The main reason is that human power is limited, and bicycles need to be very efficient to be fun to ride. Bicycles have to be light, yet strong. All parts must be designed for minimum resistance, yet they need to function with minimal maintenance. And since a bicycle can reach very high speeds on downhills, bicycles also need to be safe. All this requires precision in engineering and manufacture, and high-grade materials. Looking at it another way, you can buy a decent bicycle for less than 10% of the price of the cheapest new car.

If your budget allows a little more, you'll get parts that function more smoothly and last longer. It's more fun to ride a bike that shifts with an effortless click than one that requires you to push hard on the lever before the new gear engages with a *Ka-chunk*. The brakes will be more powerful, and – less importantly – the bike will be lighter, too.

Spend even more and you'll get a better frame. The frame won't just weigh less, but usually it will also offer a more spirited performance. Test-ride bikes at a range of quality levels and see how they feel different. And remember: Great bikes are so much fun that almost nobody regrets the extra money they spend on a better model.

Why pro racers ride 25 mm tires

Professional racers use much narrower tires than the 38 to 55 mm I recommend. Most of them ride 25 or 28 mm-wide tires. Why? European races usually go over smooth roads, and professional cyclists are light, so a 25 mm tire is a fine choice. Bicycle Quarterly's testing found that 25 mm-wide tires are faster than narrower ones, but the performance gains level off after that. Wider than 25 mm isn't faster on smooth roads – but it isn't slower, either.

Races often are won by just a few centimeters, so it makes sense for racers to use the lightest (and thus narrowest) tires that still roll fast. For the rest of us, a few centimeters at the end of a long sprint make little difference. For us, wider tires offer as much speed as narrow ones, plus more comfort, durability and safety.

Part 2

The Science of Performance

A faster bike is more fun. It entices you to pedal harder and to go further. It allows you to envision adventures beyond the horizon; adventures that wouldn't be possible, or at least not as much fun, on a lesser machine.

Performance is not about winning races or dropping your friends on a climb. It's about becoming one with your bike, about gliding effortlessly along. It's about everything that makes cycling so special.

"We rode like birds, shifting positions without regard to the hairpin turns, pedaling hard to exhilarating speeds. When the road straightened out and I sat up to spin easily again, I had a light, happy feeling in my chest. I was in love."

The Magic Equation

"More power or less resistance – the result is the same."

The performance of a bike is defined by two elements. The first is to use the rider's power as efficiently as possible. This means reducing all resistances to a minimum: Wind resistance, rolling resistance and suspension losses, climbing resistance, bearing friction…

The second aspect is often overlooked: The bike should allow the rider – you – to develop as much power as possible. For example, if your riding position is uncomfortable, if your breathing is restricted, or if you bounce around on rough roads, your power output will be reduced. If the frame isn't in sync with your pedal strokes, your legs will hurt and limit your performance. Bicycle Quarterly's research has shown that a rider's power output can vary significantly – by more than 10% – from bike to bike.

If we write this as an equation, we get:

Power Output – Resistance = Speed

To get more speed, we can reduce the resistance, or we can increase the power output. Ideally, we'll do both.

Resistance on a bike is caused mainly by aerodynamic resistance, rolling resistance, as well as bearing and other mechanical resistance. The weight of the bike also adds resistance when accelerating or climbing.

Wind resistance is largest, but also hardest to reduce. Rolling resistance offers the greatest potential for improvement. Bearing resistance is so small that there is little potential for gains. Weight gains are small as far as the bike is concerned – the rider makes up the biggest portion of the system's weight.

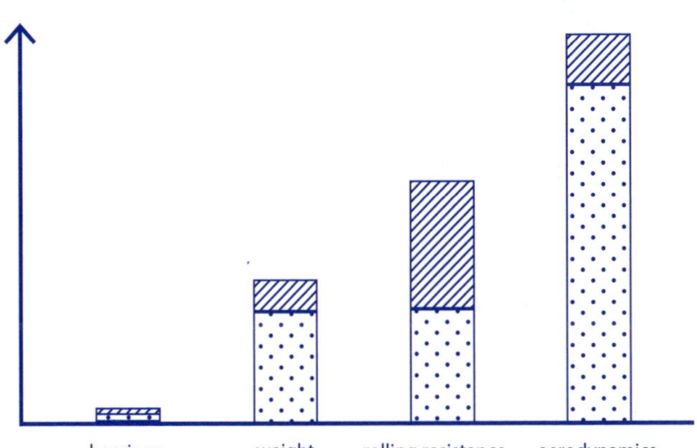

Resistances

The main resistances on a bike. Shaded areas show potential improvement. Rolling resistance has the largest potential for improvement.

Aerodynamic Resistance

"The racer-types haven't counted on the efficiency of the aero tuck. They seem surprised when, during a short descent, two randonneur bikes pass them."

Once a cyclist reaches speeds of at least 20 km/h (12 mph), wind resistance becomes the most important resistance. Aerodynamic resistance goes up by the square of the speed. To double our speed, we have to overcome four times as much wind resistance. Putting out four times as much power is difficult, so cyclists try to decrease their wind resistance as much as possible. Unfortunately, of all the resistances incurred while cycling, wind resistance is the most difficult to reduce.

Bicycle Quarterly spent three days in the wind tunnel of the University of Washington's Department of Aeronautics & Astronautics to test and optimize the wind resistance of 'real-world' bicycles.[1, 2] Wind resistance can be divided into two factors:

Wind resistance = A × c_d

A: frontal area
c_d: drag coefficient

Drag Coefficient

The drag coefficient c_d is determined by the shape of the object. The drag coefficient of a cyclist is about three times that of a modern car.

The more sharp edges, gaps and protrusions a shape has, the more turbulence it creates and the less aerodynamic it is. Unfortunately (this word pops up often in this chapter!), a cyclist's shape is not very aerodynamic, and there is little we can do to improve it. Even a set of aero wheels reduces the wind resistance of bike-and-rider by only 2-3%.

The shape of a cyclist is not very aerodynamic compared to a modern car.

Frontal Area

There is more we can do about frontal area. A small body has less wind resistance than a large one. Reducing a cyclist's frontal area is the best way to reduce their aerodynamic resistance.

In the illustration below, the cyclist on the left rides on the hoods in a relatively upright position. The cyclist in the middle rides in the drops in a more inclined position – their frontal area is 6% smaller. The cyclist on the right crouches in a full aero tuck that is not only lower, but also narrower, since the arms are pulled inward. This reduces the frontal area by 33% compared to the cyclist on the left. An additional 5% gain in the overall wind resistance is the result of the rider's much-tighter 'packaging' that reduces the drag coefficient.

Three riding positions (from left): On the hoods, in the drops, full aero tuck.

Riding Position

The biggest wind drag is caused by the rider, since the rider has the greatest frontal area. For example, lowering the stem by 20 mm (0.8") decreases the wind resistance by 5% – about twice as much as a set of aero wheels. However, unlike the aero wheels, stem height must match the rider's power output. Lowering the stem too much can make the bike less comfortable and thus reduce the rider's power output. This would negate any gains from being more aerodynamic.

Aero bars improve a cyclist's aerodynamics more than almost anything else, because they move the rider's arms inward until they are in-line with the torso. This decreases the rider's frontal area, while also improving the drag coefficient by closing the cavity underneath the rider's chest. Aerobars work best when set up very narrow, but here again, we get into issues of comfort and power output.

When descending, adopting a full aero tuck is even more aerodynamic than aero bars. The rider grasps the handlebars next to the stem, tucks in their elbows until they touch the top tube, and lowers their chin until it almost touches their hands.

The full aero tuck: Hands hold the bars next to the stem, chin is just above the hands, elbows and knees are pulled inward and touch the top tube.

Thus we get the previously mentioned 33% reduction in frontal area, and since the aero tuck closes the cavity under the rider's chest, it also makes the shape more aerodynamic. A rider cannot pedal in this position, but even on moderately steep slopes, the reduction in wind resistance is greater than the gains obtainable from pedaling hard.

Pedaling down a 6.5% hill requires 400 watts to keep up with a rider coasting in the aero tuck. That is close to the maximum of a trained cyclist, and 6.5% isn't even that steep. It's much less fatiguing to tuck and coast than to work hard and pedal.

Recently, professional racers have adopted the super tuck: They slide forward off the saddle and crouch on the top tube. Compared to the classic aero tuck, this lowers their backs. However, they cannot tuck their arms inward, but instead hold the handlebars in the drops. Which tuck is more effective? The super tuck is probably more aero on bikes with very low stems, but sliding off the saddle makes the bike less stable – hence the UCI has proposed banning the super tuck. The traditional aero tuck provides similar gains with less risk.

On downhills steeper than 6.5%, coasting in the aero tuck is faster than pedaling with 400 watts.

Clothing

Clothing is the easiest way to improve a cyclist's aerodynamics. A loose-fitting cycling jacket increases the bike-and-rider's wind resistance by a whopping 12%. A tighter-fitting jacket causes a much smaller increase in drag of 4%. For a rainy event, a good cycling jacket may offer greater performance benefits than a set of aero wheels.

Once you wear relatively tight-fitting clothes, further improvements are harder to achieve. Swapping a long-sleeve jersey for a short-sleeve one reduces your drag by less than 1%. A good guide to the wind resistance of your clothes is whether they flap in the wind during fast descents. If they do, consider changing them for closer-fitting ones.

Tire Width

Wide tires create only a negligible increase in frontal area. We tested a 24.5 mm-wide tire and a 31 mm-wide tire, with similar tread patterns. Both with and without fenders, the raw measurements showed a 1% lower wind resistance for the narrower tire, but the results were not statistically significant. This means that the differences in wind resistance between the two tire sizes are so small that we could not determine whether they are real differences or simply caused by 'noise' – small variations in the setup that inevitably occur during testing. In other words: Any aerodynamic differences between narrow and wide tires are too small to be important in real-world riding.

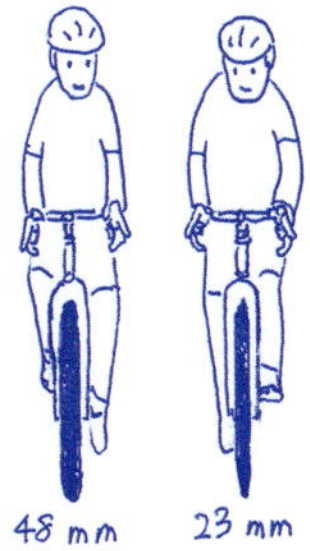

A 48 mm tire increases the frontal area by only 1.5% compared to a 23 mm tire.

Fenders

Racing bikes don't have fenders… so it's natural to think that fenders will make a bike slower. If it's not the added weight, then they must be less aerodynamic.

The reality is more complex. The forward extension of a front fender – if it follows the curve of the tire and extends beyond the crown of the wheel – acts as a fairing that shields the top of the tire from the wind. Since the top of the tire moves forward at twice the speed of the bike, this has a small, but significant effect, reducing the bike-and-rider's wind resistance by 1% – or about half as much as a set of aero wheels. For that reason, racing motorbikes have small front fenders. Racing bicycles aren't allowed to have aerodynamic fairings, otherwise, they might have fenders, too.

The portion of the fender behind the fork crown adds a little turbulence and drag. This almost cancels out the improvement you gain from the front portion, but a good front fender has slightly less drag than no fender at all. We did not measure the rear fender, but it is likely that a similar effect occurs. Mudflaps increase the drag, but the overall effect of fenders, even with a small front mudflap, is a slight improvement in wind resistance.

The front portion of each fender shields the tire. The rear portion catches the wind.

Bags

Most bags increase the wind resistance of the bike. Front bags are more efficient than rear bags. A teardrop shape is the aerodynamic ideal: wide at the front, narrow at the rear. A front bag fits within that shape, a rear bag sticks out on the sides.

Handlebar bags create the least drag, about 2-5% for a standard Berthoud model with side pockets. The lower the rider gets behind the bag, the smaller is the increase in wind resistance: The bag starts to act as a fairing. This suggests that a bag could be designed that actually reduces wind resistance, at least in a straight-on headwind.

Saddlebags that extend beyond the rider's thighs significantly increase the wind resistance drag: about 7% in our testing. Panniers aren't as unaerodynamic as they look. Front panniers increase drag by less than 10%; rear panniers by about 15%. In crosswinds, front panniers shield the front spokes, and the drag increase is reduced to just 3%. Rear panniers become much worse in crosswinds, increasing the wind resistance of bike and rider by a whopping 30%.

We did not measure bikepacking bags in our testing. Frame bags should not increase drag, and underseat bags could act as fairings and improve aerodynamics slightly, if they do not extend too far back.

Handlebar bags don't increase drag by much. Front panniers are more aero than rear panniers, especially in crosswinds.

Fairings

Wouldn't it be nice to use a fairing and reduce the drag of the cyclist? Velomobiles – fully faired, teardrop-shaped recumbents – have remarkably low wind resistance. On standard bikes, most fairings increase the rider's drag. Instead of smoothing the airflow, they simply are additional shapes that the rider must push through the air. We tested two custom-made fairings in the wind tunnel, and both caused significant increases in drag in most riding positions.

One exception is the aero tuck position: When the rider placed their head right on top of a small fairing, wind resistance was slightly less than without the fairing. Fairings must reach very close to the rider's body to be effective – like they do on racing motorbikes. On a bicycle, this is difficult to achieve since cyclists change their riding positions, so a fairing that offers benefits in one position will not work in others.

Fairings add drag (left) unless they are very close to the rider's body (right).

Combined Effect

The combined aerodynamic effect of all these factors can be very significant. We tested four riders in the wind tunnel: a track pursuit racer on a full aero bike with aerobars and a tight-fitting skinsuit; a racer; a randonneur; and a city rider dressed in street clothing. As expected, the track rider on the aero bike posted the lowest aerodynamic resistance. The randonneur and racer both had about 36% more wind resistance than the track rider. The commuter's wind resistance was a full 80% greater than that of the track rider. The increases in wind resistance are almost entirely due to riding position and clothing.

Wind resistance varies with riding position and clothing.

Drafting

The easiest way to reduce your wind resistance is by drafting another rider. In our study, we had a tall rider (185 cm/6'1") draft a short rider (155 cm/5'1"). Even in this 'worst-case' scenario, the taller rear rider's drag decreased by 38% – roughly the same as the full aero tuck. If the front rider is the same size as the rear rider, the reduction in drag should be even more significant.

Perhaps more surprising, drafting decreased the drag of the leading rider – by 3% in our test. The front rider receives a push from the rear rider. This effect is more pronounced in larger groups. Riding at the front of a peloton is much easier than riding alone, which is why solo break-aways in races rarely succeed. When you stand by the roadside during a race, the air wave generated by the large group of racers in the peloton will hit you several seconds before the first rider arrives. It's like a big truck passing you on a highway.

Drafting helps not only the rear rider. The front rider receives a push, too.

Aerodynamics Summary

Wind resistance is large, but difficult to reduce. Beyond putting your bags on the front, don't worry much about your bike, but work on your position instead: Get as low and narrow as is comfortable, wear tight-fitting clothing, and ride with friends, so you can take turns drafting and lowering each other's wind resistance by a third or more. And during downhills, coast and get in the aero tuck for minimum resistance and maximum speed.

Rolling Resistance

"As I enter the rough road, the bike's front wheel, handlebars, computer and my hands become a blur as they shake violently. To keep the bike moving, I have to pedal harder. A lot harder."

Reducing rolling resistance is how you can make the largest improvement in your bike's performance. At moderate speeds, a fast tire can make your bike 17% faster than a slow tire.[3] That is a much greater improvement than any other factor that affects the speed of your bike. Even at high speeds, the effects of rolling resistance are surprisingly large. Unlike wind resistance, rolling resistance is relatively easy to reduce. Faster tires are more comfortable, too, creating a win-win situation. The only downside is that they tend to get more flats, but as riders are adopting wider tires that run at lower pressures, flats actually become much less of a problem: Wider tires roll over debris that would puncture a narrower, harder tire.

Hysteretic Losses

Under the weight of the rider, the tire deforms where it touches the road. As the wheel rotates, the tire continuously deforms at the front end of the contact patch, and it springs back at the back as it leaves the ground. This deformation requires energy – imagine squeezing a tennis ball in your hands.

To decrease these hysteretic losses, we have two options. First, we can make the tire easier to deform, so it absorbs less energy, by using a thin, supple casing that absorbs less energy as it deforms. It's like squeezing a soft marshmallow instead of a stiff tennis ball.

Second, we can try to reduce the deformation of the tire, by increasing the air pressure. In theory, at infinite air pressure, the tire casing no longer deforms at all. No deformation means no hysteretic losses, so in the past, this was thought to reduce rolling resistance almost to zero.

Today, we know that higher tire pressures do not make a bike faster.[4] Inflating tires harder does reduce hysteretic losses, but it increases suspension losses, and the two effects cancel each other.

Rolling resistance consists of two parts:

- Hysteretic losses: energy lost as the tire deforms at the bottom of the contact patch (right)
- Suspension losses: energy lost as bike and rider vibrate (following pages)

Suspension Losses

Suspension losses are caused by internal friction as bike and rider vibrate. This friction occurs mostly in the rider's body. As the body vibrates, the tissues rub against each other. This friction uses energy. Energy on a bike can come only from one source: the rider's power output. If you lose energy somewhere, the bike slows down.

What we feel as discomfort when our bodies vibrate actually is lost energy. This has been known for decades: In a study of tank seats, the U.S. Army found that suspension losses can amount to 2000 watts before the vibration becomes so intense that the discomfort is unbearable.[5] The Army study also found that discomfort and energy loss are directly proportional: The more discomfort, the greater the suspension losses. However, this research has only recently been applied to bicycles.

We measured the suspension losses on a bicycle by riding on two adjacent surfaces: very smooth, new pavement; and a 'rumble strip' that provides standardized small bumps.[6, 7] On the rumble strip, we went at the same speed as on the smooth pavement.

Suspension losses are caused by internal friction as bike and rider vibrate.

Nothing except the road surface changed, yet we needed more than twice as much power to keep the bike moving at 25 km/h (16 mph). The extra power – 290 watts – was lost to the additional suspension losses on the (very) rough surface.

Rumble strips are an extreme situation, but our experiment shows how important suspension losses really are. 290 watts is huge – more than even a strong rider can put out for long distances. Even on smooth roads, suspension losses are significant. They are the reason that a supple racing tire rolls as fast at 5 bar (75 psi) as it does at 6.5 bar (95 psi). The decrease in hysteretic losses at the higher pressure is canceled entirely by increased suspension losses. Other tests showed that even at much higher pressures of 13 bar (190 psi), the tire doesn't roll faster.

On rough roads, suspension losses increase, and higher tire pressures make the bike slower. During our rumble strip experiment, the same 26 mm tires consumed 20% more power at 6.6 bar (95 psi) than they did at 5.2 bar (75 psi). Gravel racers have known this for a long time: When the surface is really rough, the lowest pressure you can run safely is also the fastest.

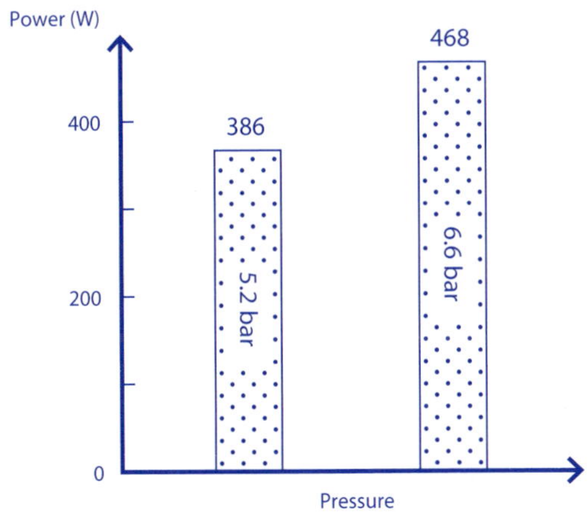

Rumble Strip

On rough roads, higher pressure makes your tires slower. (Data for 26 mm tire at 19.3 mph/31 km/h).

Tire Pressure Summary

- Higher pressure decreases the hysteretic losses in the tire.
- Higher pressure also increases vibration and thus suspension losses.
- On smooth roads, the two factors cancel each other, and higher pressures neither increase nor decrease your speed.
- On rough roads, suspension losses are more important, and most tires roll faster at lower pressures.

Inflating your tires harder does not make your bike faster. In fact, depending on the road surface, it can increase resistance and slow you down! To obtain the best performance on a variety of surfaces, run your tires at relatively low pressures.

Experimentation is the best way to determine the optimum tire pressure. The pressure written on a tire's sidewall is only the maximum safe pressure – it's not the pressure you should use.

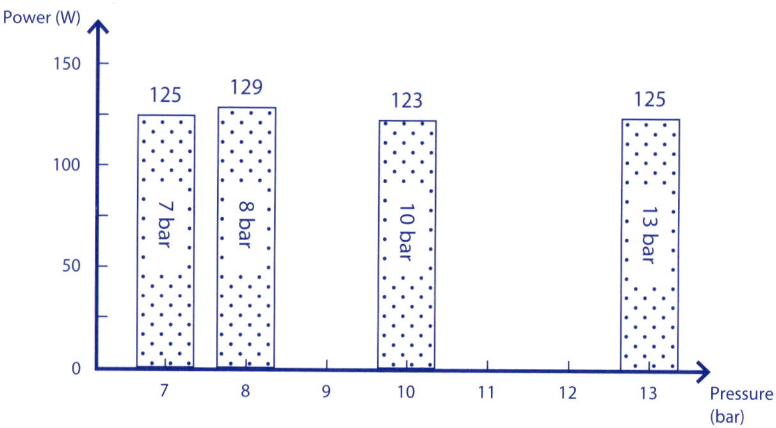

On smooth roads, increasing tire pressure from 7 to 13 bar (100 to 190 psi) does not change the power to pedal the bike. (32 mm tires; 15.5 mph/25.0 km/h)

Placebo of High Tire Pressures

Many riders are convinced that inflating their tires harder makes them roll faster. "We can feel it," they say. But we cannot feel speed. What we feel is the vibration of the bike. As speed increases, our tires hit more road irregularities per second, and the bike vibrates with higher frequencies. We are conditioned to equate these higher vibration frequencies with greater speed.

Inflating our tires harder is another way to increase the frequency of the tires' vibration. In other words: Higher tire pressures make the bike *vibrate* faster without actually *going* any faster. This has tricked generations of riders into believing that pumping up their tires harder makes them faster. It doesn't!

This is also one reason why wide tires feel slower to many riders. Wider tires roll at lower pressures and transmit less vibration. They give the rider the same sensations as riding slower, even though all tests show that, as long as the tires have the same casings, wider tires at lower pressure don't roll any slower than narrow, high-pressure tires.

It takes time to un-learn the reflex that equates vibration with speed. When you ride with others, you'll notice that you have no trouble keeping up even if you have a slow leak in your tire, until it's almost completely flat. And if you ride in timed events, you'll find that running wider tires at lower pressures won't hurt your speed.

Supple Casings

The best way to make a tire faster is to use a supple casing.[8] Supple casings not only reduce hysteretic losses (they are easier to flex), but also reduce suspension losses (they transmit less vibration). This is a win-win situation, and the effects are dramatic: Changing tires – even among high-performance models from different brands – can increase your speed on the bike by 8-20%. This is by far the greatest change you can make to the performance of your bike.

We ran our tire tests at moderately high speeds, between 24 and 30 km/h (15 and 19 mph). At lower speeds, the advantages of supple tires are even more pronounced, because wind resistance is reduced and rolling resistance makes up a bigger percentage of the overall resistance. At higher speeds, aerodynamics are more important, but supple tires still make a big difference. That is why all professional racers ride on supple tires.

In Part 3 of this book, we'll look at how to optimize tires – their casings, width and pressure – to optimize performance, comfort and safety.

Supple tires can make a very significant difference in your bike's speed.

Wheel Size

At first sight, it appears that larger wheels should roll faster. By making the wheel larger, you make bumps smaller in relation to the wheel, and everybody knows that it's easier to go over smaller bumps than over bigger bumps. And yet our testing found that wheel size does not affect how fast a tire rolls. Both on smooth pavement and on the bumpy surface of rumble strips, 559 mm (26") tires rolled as fast as 584 mm (650B) and 622 mm (700C) ones.[9]

Why don't larger wheels roll faster? Bicycle wheels aren't round, but flat at the bottom. Small obstacles are absorbed by flex in the tires, not by lifting the entire bicycle over them. How fast the wheels roll is determined by how much energy the tires consume as they go from round to flat, and back to round again, as the wheels rotate. Wheel size doesn't matter as long as the bumps are small enough to be absorbed by the tires.

If you want to absorb bigger bumps, the first step is to make the tire taller. Usually this means going to a *smaller* rim size. For example, a 650B x 48 mm tire has the same outer diameter as a 700C x 28, but it can absorb much-larger bumps. If the bumps are so large that they cannot be absorbed by the tire, a larger wheel may roll faster over them.

For road riding, including gravel roads, wheel size does not affect a tire's performance. However, wheel size does affect the bike's handling (see p. 87).

Bicycle wheels are flat at the bottom, and road irregularities are absorbed by the tire flexing.

Other Suspension Losses

Anything that reduces vibration will increase your bike's speed. A stiff fork will require more power than a suspension fork. You can get the same effect from a steel fork with small-diameter, flexible blades. On rough surfaces, a fork that reduces vibration improves your speed.[10]

Shock-absorbing forks can make your bike faster on rough surfaces. (Power data for 19.3 mph/31 km/h)

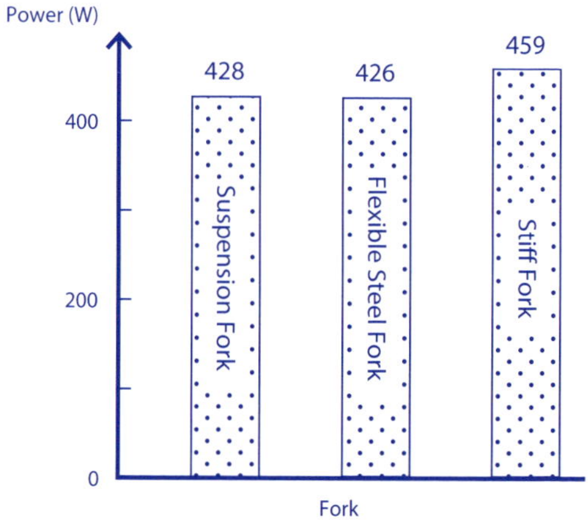

Rough Road

Damping

How can a supple tire absorb all that vibration without consuming much energy, when the same vibration causes large energy loss when it's absorbed in the rider's body? The difference is in the damping.

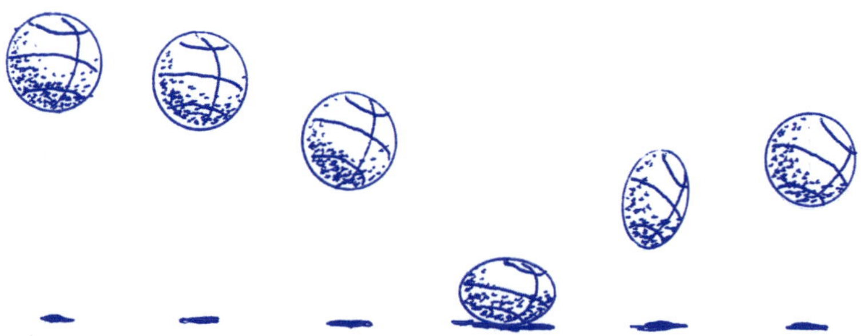

A basketball has almost no damping.

Damping is how movement is stopped and converted into heat. When you drop a basketball, it bounces back almost as high as before, because it absorbs very little energy as it bounces. There is almost no loss: the energy of the impact is stored briefly in the ball and then released to accelerate it upward again.

When you drop a beanbag, it comes to rest on the ground without bouncing. All the kinetic energy has been absorbed through friction between the beans that fill the bag.

The basketball is undamped, while the bean bag has very effective damping. The basketball won't get hot no matter how hard you bounce it, but the temperature of the bean bag increases with each drop, as energy is transformed into heat.

The more supple a bicycle tire is, the more it acts like a pure air spring. It deforms as it encounters surface irregularities, storing the energy in the deformed casing. As the tire rolls off the back side of the bump, the tire pressure pushes against the bump, thus releasing the energy and accelerating the bike again. A stiffer, heavier tire has more damping. It absorbs more energy and doesn't spring back as forcefully.

Like the bean bag, the human body is very effective at damping. Vibration that reaches our bodies slows us down. The secret is to have the undamped tire deal with road vibration instead of the highly damped rider's body. It uses less energy, and it's more comfortable.

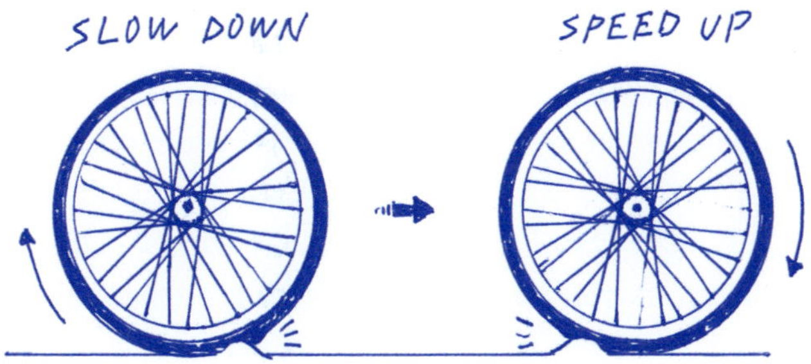

A supple tire acts like an air spring. It compresses when hitting a bump, then pushes off the other side.

Weight

"Our test bike weighed just under 17 pounds with wide tires, a leather saddle, three bottle cages and a pump. Just standing in the hallway, the orange carbon fiber machine was quite something."

Weight is often considered the holy grail of bicycle design. It's easy to measure – all you need is a scale. Simple physics tell you that a lighter bicycle goes uphill faster. This is easy to understand, but the big question is: How much faster?

As with aerodynamics, it's a mistake to look at a bike without considering the rider. A 9 kg bicycle is one kilogram lighter than a 10 kg bike. And yet it would be a mistake to think that the difference amounts to 10%, because this overlooks the rider. Even with a featherweight 60 kg professional racer on board, the total weight is 70 kg. Now the 1 kg weight difference is just 1.4%.

Our racer will go up a steep hill about 1% faster on the lighter bike (wind resistance increases with speed, so the speed increase isn't the full 1.4%). One percent means six seconds over a ten-minute climb. If our racer is barely hanging with a pack, six seconds is the difference between making it over the top with the group or getting dropped. In that case, 1 kg off the bike can be worth a lot.

For most riders, six seconds over a ten-minute climb will not be noticeable. And even for the racer, other factors, like rolling resistance or the frame's flex characteristics, are more important. That is why racers all ride on supple, handmade tubular tires. You may be surprised that many bikes in the Tour de France are heavier than the required minimum weight: Pros choose reliability over lightness. Even for them, a few grams make no difference.

A racer who is a just few seconds slower on a long climb gets dropped.

Wheel Weight

What about lighter wheels? Do they climb faster, or at least accelerate faster? As far as acceleration is concerned, the difference is small, simply because cyclists don't accelerate very fast to begin with. Even a small economy car has a power to weight ratio that is five times greater than that of the strongest professional racer. That is why wheel weight has only a small effect on acceleration. This goes against conventional knowledge, but it's easy to calculate.

Two professional sprinters, each putting out 1600 watts, start their sprint side-by-side at a speed of 50 km/h (32 mph). One has superlight wheels (1430 g), the other relatively heavy wheels (2000 g). After a 200 m all-out sprint, the racer on the lighter wheels is just 29 cm (11 in) ahead.[11] The lighter wheels do accelerate faster, but the difference is smaller than most cyclists think.

If lighter wheels really did 'spin up' faster, professional racers could gain a significant advantage by using smaller (and thus lighter) wheels. Current rules allow wheels as small

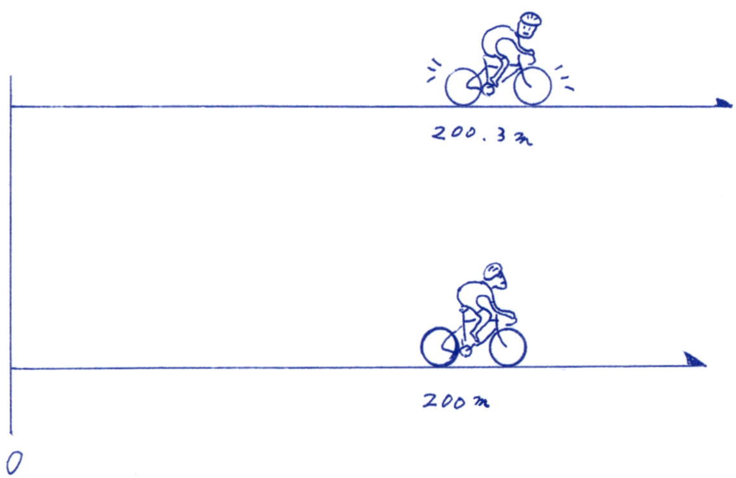

Wheel weight makes only a small difference in how fast a bike accelerates.

as 55 cm in diameter, yet all the pros ride wheels that are 10 cm larger than the minimum. (A 700C x 25 mm wheel has a diameter of about 65 cm.) Numerous racers and builders have tried smaller wheels in the past, but apparently none have found significant advantages despite their lighter weight. Even track sprinters use big 700C wheels.

What about climbing? In theory, at constant speed, it should not matter whether a bike's weight is in the rotating wheels or on the parts that move in a linear fashion (frame, fork, handlebars, etc.). However, when we look at 'planing' and how the frame flex characteristics affect the rider's power output (see p. 60), we'll see that the interactions between rider and bike are complex. It is conceivable that lighter wheels allow micro-accelerations of the bike with each pedal stroke, thus reducing the resistance to the pedal stroke. Reducing that resistance could allow the rider to put out more power – the bike pushes back less against the rider's legs.

In the absence of controlled studies, this remains purely hypothetical. If lighter wheels made bikes climb faster, then small-wheeled bikes would offer an advantage. The fact that small-wheeled bikes aren't known to climb faster suggests that any benefit from 'micro-accelerations' during climbing may be too small to make a difference. The best climbers in the Tour de France race on large 700C wheels...

Rocking the Bike

Lighter wheels feel different when you rock the bike side-to-side while sprinting out of the saddle. At a cadence of 90 rpm, the bike changes direction, from right to left or vice versa, three times per second. The lateral accelerations are quite large. Lighter wheels make the bike easier to rock. This is something you can feel, and it may contribute to the impression that lighter wheels are faster. In fact, it may even make the bike faster, since less energy is needed to rock the bike, and thus the rider can use more energy to propel the bike forward.

It's possible that there is an optimum rotational inertia in the wheels that allows riders to maximize their power output as they rock the bike: too little resistance, and there is nothing to push against; too much, and the bike resists the rocking motion. In that case, the rotational inertia of wheels should be optimized, rather than be reduced as much as possible.

Over the last 80 years, the wheel size and wheel weight of racing bikes hasn't changed much – perhaps because they represent an optimum for the best sprinting performance. For the same reason, it makes sense to reduce the wheel size (rim diameter) of all-road bikes with wide tires, to keep the rotational inertia – and thus the out-of-the-saddle feel – at the same optimum level.

The weight of the entire bike also affects how it feels when you rock the bike while riding out of the saddle. A heavy handlebar bag, even a set of fenders, or how high the water bottles are placed on the bike, change the feel of the bike. Panniers mounted on low-riders are closer to the bike's fulcrum and inhibit the rocking of the bike far less than higher-mounted loads. The out-of-the-saddle feel is probably the biggest difference between a svelte racing machine and a bike equipped for touring or adventures.

Out of the saddle, the weight affects how easy it is to rock the bike.

Cyclocross and Portaging

Bike weight matters much more when you are carrying the bike. In that case, 1 kg off a 10 kg bike really makes the bike 10% lighter. In cyclocross, a lightweight bike is useful because you have to lift the bike every time you jump across a barrier or run up a hill that is too steep to ride. A travel bike that you carry should also be as light as possible.

Weight Summary

The importance of bicycle weight is often overestimated. Beyond the pure physics, there is beauty in a lightweight bike, and I firmly believe that a bike should not be heavier than necessary. Also, the qualities that make tires and frames perform better also usually make them lighter, which is why many lighter bikes really do offer better performance. However, chasing light weight to the detriment of other qualities does not usually result in a bike that is faster or more pleasant to use.

Weight makes a difference when portaging the bike.

Bearing and Mechanical Resistance

"Ceramic bearings offer few performance benefits, because traditional ball bearings are so efficient. One half of almost zero is still almost zero."

Ball bearings made bicycles possible. They spin with little resistance, they don't wobble, and they last a long time. We take them for granted, but without them, your bike would neither roll easily nor steer with precision.

There is not much to say about bearing resistance. A few years ago, ceramic bearings were considered the best performance upgrade for your bike. However, even low-quality bearings are incredibly efficient, and you don't gain much from better bearings.[12] Ceramic bearings may have half as much resistance as standard bearings, but the gains are minimal. When the resistance of standard bearings is almost zero, significant further reductions are impossible.

The same applies to rubber seals that protect bearings from contamination. They make bearings harder to spin with your hands. On the road, the added resistance is too small to make a difference.

Quality bearings tend to last longer. Even badly worn bearings have little friction, but when they pack up completely, their resistance becomes infinite, and your ride is over!

The resistance of the drivetrain is also small. Even severe cross-chaining, as is common on modern One-By drivetrains, does not increase the mechanical resistance by much. What does make a difference is the rear cog size: The chain has more resistance on cogs smaller than 14 teeth.[13] Derailleur pulleys usually are smaller than that, but the chain doesn't run over them under load, so the possible gains from extra-large pulleys are small. Even a dirty chain does not significantly increase your drivetrain's resistance.

Overall, the resistance of bearings and drivetrain is too small to have a significant impact on the performance of the bike. High-quality bearings and a smooth drivetrain make a bike feel nicer, though. With poor bearings or an inefficient drivetrain, you may feel a grittiness through the pedals. This can detract from the enjoyment of cycling.

Bearings make your wheels and cranks spin and your headset turn.

Frame Flex

"I got onto his wheel, and when he slowed slightly, I went by. I pushed hard, enjoying the response of the bike and the intensity of the effort. I was surprised at the size of the gap when I reached the crest."

So far, we have covered the resistances that slow a bike. Now we look at the other side of the equation: the rider's power output. In the past, it was often assumed that a rider's power output was independent of the bike he or she rode. Today, we know that this is not the case.

On a most basic level, an uncomfortable bike won't allow its rider to put out their maximum power. A tired rider has less power, so a bike that fatigues its rider prematurely will be slower. But there is more to it: Even a fresh rider's power output varies on different bikes. Much of this relates to the flex in the frame. In the past, it was often assumed that stiffer frames are more efficient. The reality is more complex.

Despite many attempts, nobody has been able to show that stiffer frames are more efficient at transmitting the rider's pedaling power. Racers always have preferred lighter frames, even though they are more flexible: less material means less stiffness. Even during flat races, where the small weight differences between heavy and light frames are of little importance, racers prefer lightweight – and less stiff – frames.

This was most pronounced with classic steel frames. Cheap frames made from thickwall tubing were much stiffer than high-end frames made from lightweight, butted tubing. Yet no racer ever preferred the cheaper, stiffer frames. A myth developed to explain this: It was thought that high-end steels were stiffer. This is not true. All steels have the same stiffness.

Why do racers prefer frames that aren't very stiff? The right amount of flex in the right places actually makes a bike faster, because it increases the power you can put out. Grant Petersen of Rivendell once compared it to jumping on a sprung floor: Your strength is the same, yet you can jump higher and with less fatigue on a sprung floor than on unyielding concrete. This is a useful analogy. The sprung floor is flexible, but the flex must be in tune with the jumping gymnast. A foam pit is even more flexible, yet you cannot jump at all in the foam pit. Just the right amount of flex in the floor allows the highest jump. The same applies to bicycle frames.

Optimizing frame flex can increase a rider's power output by 12%.

Japanese Keirin track racers have known this for many years.[14] The careers of these racers often span 20 or more years, and their bikes' technology is firmly limited by the rules of the races. On average, Keirin racers buy two new frames every season, so they are able to experience dozens of similar bikes throughout their careers – with the only change being in how much the frame flexes. Japanese framebuilders say that the hardest part of making a Keirin frame is to tune the frame's flex characteristics to the rider's pedal stroke and power output.

Planing

Frame flex stores energy like a spring and returns it to the drivetrain when released. Depending on where the flex occurs, the return can help the rider's pedal stroke in moving the bike forward or work against it. On a frame that flexes just right, a little 'give' enables the rider to push harder on the pedals during the down stroke (where most power is generated). Excess force is stored as frame flex and released during the dead spots of the pedal stroke. When riders get in sync with their frames, they can put out more power with less fatigue.

A frame with optimized flex is not just faster, it also feels different. Pedaling harder increases the rider's heart rate to the max, but their legs don't hurt. The bike allows the rider to pedal harder and go faster. The frame's response to the pedal strokes feels like a boat rising out of the water, going faster with only a little extra energy input.

"You mean, it planes," said Matthew Grimm of Kogswell Cycles, when I described the phenomenon to him. Deciding that the phenomenon needed a name, we've used this term to describe it.

Very rigid frames tend to resist this extra energy input (like a concrete floor resists jumping). They make it harder to pedal beyond a certain speed. It feels like the bike's performance hits a brick wall. Trying to go faster feels like pushing against that wall: The wall doesn't move, yet the rider's legs fatigue. This is often felt as a burning sensation in the legs, whereas on a bike that planes, the rider's cardiovascular maximum limits the bike's speed. The rider is out of breath, but their legs don't hurt.

If the frame is too flexible, or if the flex is in the wrong places, the frame can be out of sync and work against the rider's pedal strokes. On these frames, even small increases in speed seem to require large increases in effort. Riding these bikes fast can feel like jumping in a foam pit.

The basic idea of planing has been confirmed in a series of double-blind tests.[15, 16] We used three bikes that were identical (same geometry, same paint, same components), except that one had different flex characteristics (top and down tubes with different wall thickness). The other two were the same. (Two identical bikes allowed checking whether some unknown variable – frame alignment, bearing adjustment, wheel trueness, etc. – affected the performance.)

Even though the test riders did not know which bike they were riding, Rider A was consistently faster on Bike 2.[17] His average power output was about 12% higher. Naturally, the rider's speed was also higher – by 6-8% – on Bike 2.

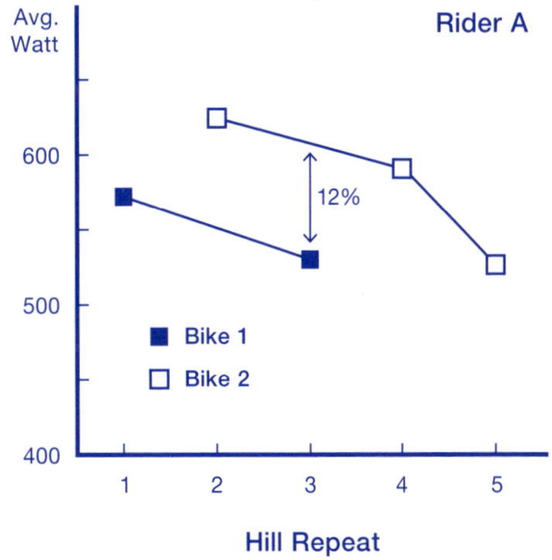

Rider A puts out about 12% more power on Bike 2 than on Bike 1. Fatigue reduces power output during Hill Repeat 5.

He reported a noticeable difference in feel, too: Bike 2 'planed' much better for him. Rider A fatigued with each subsequent run, but each time he rode the Bike 2, his power output rebounded.

The difference between Bike 1 and Bike 2 was relatively small. Both were made from standard-diameter steel tubing; Bike 2's tubing walls were 0.2 mm thinner. This alone resulted in a significantly higher power output for Rider A. The speed difference of 6-8% is much more than he would get from aero equipment or from reducing the weight of the bike.

'Planing' explains why some slightly heavier steel bikes can match or exceed the performance of lighter carbon bikes: Optimized flex characteristics allow the rider to put out more power, which has a greater effect than the bike's weight. Only when both bikes have the same flex characteristics does the weight determine which one is faster uphill.

What about energy absorbed in the frame as it flexes? The energy is returned to the drivetrain as the frame returns to its original shape. Very little energy is lost to hysteretic losses because frames are undamped springs. In theory, metal frames are better springs and absorb less energy than carbon fiber. In practice, the difference is too small to matter.

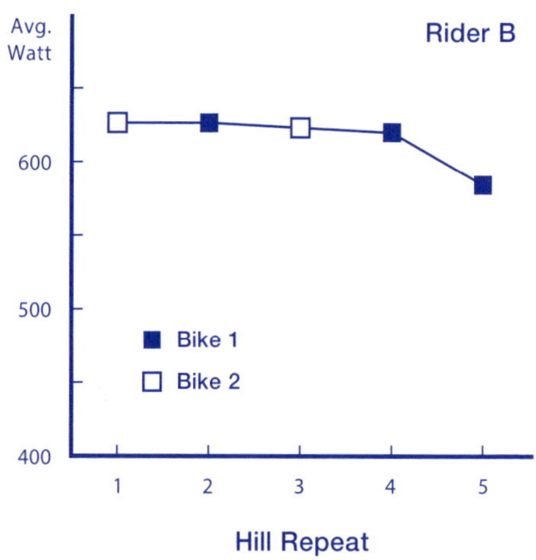

Rider B performs equally well on Bike 1 and Bike 2. Fatigue reduces power output during Hill Repeat 5.

Tubing Selection

Our tests consisted of uphill sprints, with both testers riding side-by-side. On this day, Rider B was the stronger of the two. During every race up the hill – except Repeat 2 – he pulled ahead. For him, Bike 1 and Bike 2 did not cause a difference in power output. However, Rider B also reported that the sprints felt harder when riding Bike 1 (even though he did not know which bike he was riding in this double-blind test).

For Rider A, the more flexible Bike 2 was faster. For the slightly stronger Rider B, both bikes performed equally well. For an even stronger rider, the stiffer Bike 1 might have been fastest. It appears that stronger riders perform well with stiffer frames.

During the 1950s and 1960s, top racers rode lighter and thus more flexible frames for time trials, which require long efforts at relatively moderate power outputs. They used stiffer frames for road races that are won in short bursts of acceleration at very high power outputs. Similarly, Japanese Keirin racers use more flexible frames during the early season and stiffer frames when they are at their peak form. This matches what we found in our tests.

What surprised us when we tested frame stiffness was this: Even though both testers were relatively tall (180 cm/5'11"; frame size: 57 cm) and strong riders (peak power: 700+ watts), they performed best on very flexible frames. Bike 1 was a classic standard racing frame (wall thicknesses: 0.9-0.6-0.9 mm). Bike 2 was made from superlight tubing (wall thicknesses: 0.7-0.4-0.7 mm). Even the standard frame is considered a 'noodle' by today's standards, but the 'superlight' one is made from tubing that most would consider too flexible. Yet there was no doubt: The most flexible frame worked best for Rider A, and even the stronger Rider B preferred its feel.

There wasn't any apparent downside to the 'superlight' frame. When descending, both frames felt the same. The superlight frame didn't flex noticeably as we leaned hard into corners. It just climbed better. Both testers then ordered custom all-road bikes with the same superlight frame tubes for their personal use, and these have performed fine under all conditions. They've also proven very durable.

If the most flexible frame tubes available today work well for our tall, strong testers, then many frames for smaller and/or less strong riders may be stiffer than is ideal. This suggests that, when in doubt, many riders might perform better on a more flexible frame – while maintaining the 'balance of frame stiffness' described below. That is what Andy Hampsten did when he ordered the bike he rode to victory in the 1988 Giro d'Italia. His frame was made from similar 'superlight' tubing as the test bike our riders preferred.[18]

The builders of Japanese Keirin track frames carefully tune frame stiffness to the rider's pedaling style and power output. Even these ultra-strong sprinters prefer frames that are relatively flexible, often made from standard-diameter tubing with relatively thin walls. If relatively flexible frames work best for them, they will be more than stiff enough for the rest of us.

Balance of Frame Stiffness

When we measured the stiffness of various frames, two frames tied for being stiffest: a mass-produced steel touring bike and a carbon fiber race bike.[19] And yet, despite having the same overall stiffness, they could not have felt more different on the road.

The racing bike performed great under very hard efforts. A professional racer would have loved it. Its performance only dropped off when we got tired. The touring bike did not perform nearly as well. It felt as if it was reluctant to accept our pedal power. The bike felt worse the harder we pedaled.

The difference was that the frames were stiff in different places. Their balance of stiffness was different, and the result was very noticeable on the road.

The flex characteristics of bicycle frames have changed remarkably little over the last 80 years. Most modern carbon racing bikes have a very thin, almost blade-like top tube. This is relatively flexible. The down tube is much larger in diameter and considerably stiffer. And the chainstays are as large (and stiff) as they can be, while still fitting between tires and cranks.

A classic steel racing frame has a similar balance: The top tube measures 1" in diameter and has thin walls. The down tube is larger in diameter (1 1/8") and often has thicker walls, hence it is stiffer. And many builders believe that shorter chainstays make bikes better climbers. We suspect that this isn't because they are shorter, but because they are stiffer. (A shorter tube is inherently stiffer.)

The balance of stiffness in racing frames hasn't changed in almost a century: relatively flexible top tube; stiffer down tube; very stiff chainstays. It appears that the other tubes don't have a great effect on a bike's performance.

The touring bike we tested had an oversize top tube to increase the stiffness when riding the bike with a rear load. The down tube was the same diameter, but since it's longer, it is less stiff. And the chainstays were very long and ovalized to clear wide tires. The result was a stiff top tube, slightly more flexible down tube and very flexible chainstays – a balance that is the opposite of a performance frame. On the road, the touring bike was hard to ride at speed – it was happiest if we moderated our power output and simply rolled along.

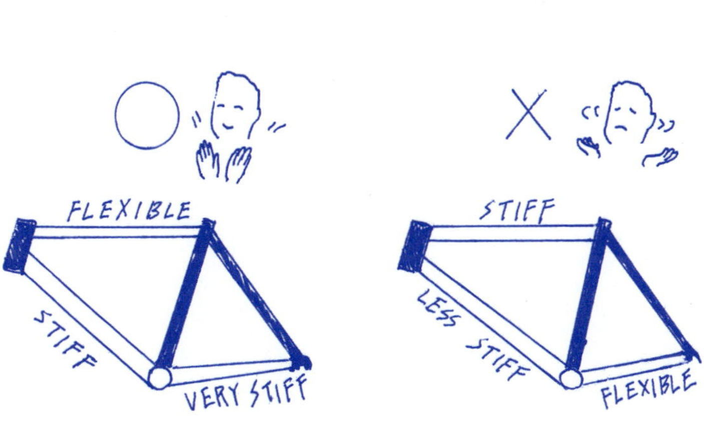

The balance of stiffness of performance frames (left) compared to some touring frames (right).

Planing Potential is not Unlimited

We've tested more than 100 bikes for Bicycle Quarterly, and we found that the fastest bikes all 'plane' similarly. They all offer the same climbing performance once their weights are equalized. Once a frame flexes in unison with the rider's pedal strokes, no further gains are possible. Sub-optimal frame stiffness appears to be a limiting factor that inhibits the rider's power output. Once a bike planes optimally, the rider can put out maximum power. There is no need to search for a frame that planes even better.

How much frame flex is best for you depends on many factors: your power output, your pedal stroke, your weight and also your personal preferences. Try as many frames as you can to figure out what works best for you.

Touring on a Performance Bike

If you want to enjoy the feel and speed of a performance bike on a tour, there is a simple solution: Pack your entire load into two large panniers and attach them to front low-riders. This lets the front wheel carry the load, and the frame of your 'touring' bike can be built with the balance of stiffness of a performance bike. In fact, many modern all-road bikes have forks that are equipped with the eyelets needed for low-rider racks, making them ideal 'fast touring bikes.'

Front low-riders allow touring on a performance bike with a relatively flexible frame.

Planing in Other Places

The beneficial flex that makes it easier for the rider to pedal harder does not always occur in the main triangle of the frame. There are other places where flex can store energy and release it to power the bike forward.

We once tested a bike with an elastomer rear suspension: The seatstays were connected to the seat tube via a rubber bumper.[20] It was sold as a more comfortable bike, but the amount of flex was very small – just a few millimeters. It was not enough to make a real difference in comfort, but the bike surprised us with its excellent climbing performance – much better than we expected from a frame made from stiff oversized aluminum. Could it be that the flex in the elastomer made the bike 'plane' for us?

The maker offered elastomers with three different hardnesses, and we machined a stiff plastic insert to ride the bike with no suspension at all. Then we repeated our testing with each of the four suspension settings. We found that without suspension, the bike climbed like we would have expected a stiff, oversize aluminum frame: Beyond a certain power output, it was difficult to pedal harder, because our legs began to hurt. The 'hard' elastomer felt the same. It was the 'medium' elastomer that made the leg pain disappear, allowing us to increase our power output until we reached the limits of our cardiovascular system. The 'soft' elastomer brought a small

Beneficial flex can occur in different places on a bike.

improvement in comfort, but our performance deteriorated. It wasn't that our legs were hurting, but we just couldn't get any power down, as if somebody was pulling on our legs as we pedaled.

The flex in the rear had to be 'just right,' otherwise it was either too small to make a difference, or so large that it had what mountain bikers call 'suspension bob,' where the frame moves completely out of sync with the rider's pedaling. Frame flex that can be tuned to match the rider's power output is an interesting idea. Unfortunately, the maker discontinued this bike.

Tire Pressure and Flex

Another way to fine-tune flex in a bike is through tire pressure. You need wide tires for this. At just the right pressure, a wide tire deflects and stores the excess energy of the pedal strokes, releasing it later. At Bicycle Quarterly, we have tested a number of very stiff all-road bikes that performed very well with the tires inflated to 'gravel' pressures. When we increased the tire pressure for a fast road ride, the performance became sluggish. Reducing the pressure again restored the wonderful performance.[21] It appears that with the right pressure, the tire sidewalls offered just the right amount of flex to allow the bike to 'plane.' You might try experimenting with tire pressures to see whether you can improve your bike's performance with a little extra 'springiness.'

Frame Flex Summary

Optimizing frame flex has the biggest potential for increasing a rider's power output. Other ways to optimize flex in a bike deserve further study. For now, an optimized frame appears to be the best way to make a bike 'plane': Our favorite bikes 'plane' well regardless of tire pressure and other factors.

Does planing work for all riders?

Our double-blind tests included a third rider. This rider could not reliably identify the different bikes. His performance did not vary noticeably on the different frames. It's possible that the third rider would have felt differences if the test frames had been more different. It's also possible that some riders do not get in sync with their frames, and frame stiffness does not affect their performance.

To determine how many riders can benefit from frames with optimized flex characteristics would require a study with hundreds or thousands of riders. Both our study and the experiences of many racers have shown that for many riders, optimizing frame stiffness can improve performance and feel of a bike. Yet others, even very strong riders, can perform well on almost any frame.

Try as many frames as you can to see if you find one that feels more in sync with your pedal strokes than the others. If they all feel the same, then frame flex may not matter to you.

Frame flex is important for some, but not all, riders.

Biomechanics

"Films of 1950s bike races flash through my mind; riders entering velodromes and then spinning their legs at incredible speeds in their sprint for the line. I used to think the films were being played at incorrect speeds. Now I know that on bikes like this one, it is easy and efficient to spin at 180 rpm."

Biomechanics are an important factor in a rider's power output. If a bike doesn't fit well, the rider can't perform at their best. An ultra-low position may be aerodynamic, but if it is not matched by the rider's power output, then their core and other muscles will have to expend considerable energy just to maintain the body in this position – energy that isn't available to propel the bike forward.

A position that is too high for the rider's power output will also limit their ability to pedal hard. Without the inclined upper body acting as a counterweight to the upward thrust of the legs, the rider will be unable to use their core muscles effectively to propel the bike.

Q factor

Q factor – the distance between the cranks – can influence a rider's power output. Research at the University of Birmingham found that a relatively narrow Q factor can increase power output by 1.5 - 3.6%.[22] The study found that trained cyclists had a stronger preference for – and put out more power with – a narrow Q factor than untrained cyclists.

All cyclists performed best on cranks with Q factors lower than those they usually rode. The trained cyclists in the study performed best with Q factors between 137 and 144 mm. Most modern racing cranks have Q factors of 146 - 150 mm. Many mountain bike and gravel cranks are significantly wider.

However, some riders report that they require relatively wide cranks to avoid knee problems.

Biomechanics Summary

Biomechanics are mostly about the rider and their interaction with the bike. They are important not only for power output and endurance, but also for injury prevention. Generally, a bike that is comfortable and can be pedaled without strain is also the most efficient and least likely to cause injury. If the bike is optimized for the rider and the rider has good form, pedaling hard taxes heart and lungs more than muscles and joints.

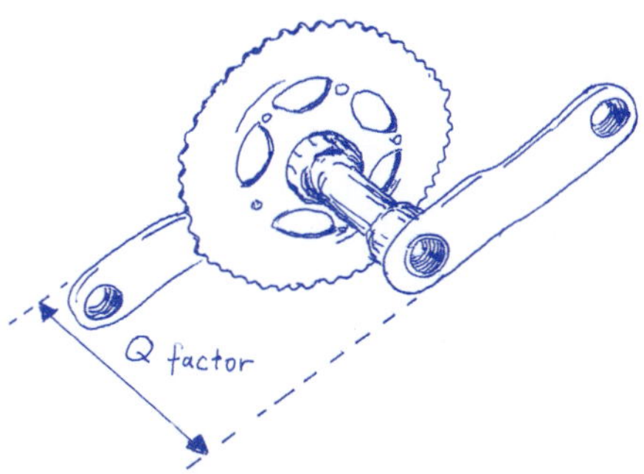

Q factor is the horizontal distance between crankarms.

Handling

"The hairpins fly toward us in quick succession. The road flows beautifully down the slopes of the volcanoes. We corner faster and faster, yet our tires' feedback indicates that we still have grip in reserve."

Everybody wants a bike that rides straight without requiring much attention, yet corners with ease. The bike should be stable and continue on a chosen line, whether on the straight or in corners, without requiring conscious input. When the rider wants to deviate from that line, the bicycle should respond immediately and with precision.

A bike should react to intuitive inputs and not require active thinking to direct it where you want to go. Such a bike is more fun and less tiring to ride than a bike that has a tendency to veer off course unexpectedly, or a bike that resists turning and runs straight whether you want to or not. Amazingly, it is possible to design a bike that does all this.

The forces that determine a bike's handling are complex. If you just want to find geometries that work well, skip ahead to page 98.

How We Balance Our Bikes

We balance our bikes by countersteering. When the bike falls to the left, we steer to the left – into the lean – to bring its wheels back under the center of gravity. A bike actually goes down the road in a series of very slight S-curves. We do the same when walking and running. It's so intuitive that we usually don't realize it.

On a bicycle, this process is greatly enhanced by a variety of mechanical factors, including the gyroscopic forces of the wheels: When the bike falls to the left, the gyroscopic forces automatically steer the front wheel to left, and bring it back underneath the center of gravity. All the rider does is guide this process so that the bike doesn't over- or undercorrect.[23]

To balance a bike that begins to fall over, we steer into the lean.

Initiating a Turn

When you corner with any vehicle, centrifugal force tries to push the vehicle to the outside of the curve. You can see this when a car corners: It leans on its suspension toward the outside of the curve. On a bicycle, leaning to the outside of the curve would make you tip over, so you have to balance the centrifugal force with another force. That force is gravity: You lean the bike into the curve until gravity pulls you into the lean as much as centrifugal force pulls you to the outside.

The faster you corner and the tighter the radius, the more you have to lean, because the centrifugal force is greater, and you need more gravity to counter it. You also need a centripetal force to 'anchor' the bike against the centrifugal force. Tire adhesion provides this force. If your tires slip, your bike loses its equilibrium, and you fall.

How do you get the bike to lean in the first place? You countersteer to move the wheels to the outside of the curve. It's the same mechanism you use for balancing the bike, except that now you unbalance the bike on purpose, so it leans into the curve. Countersteering is the only way to get a bike to lean, but there are different ways to get the bike to countersteer.

A car does not need to be balanced, so it leans to the outside of a corner. A bike is balanced even when cornering, so it has to lean into the corner.

Handlebar Pressure

The easiest and quickest way to move the wheels to the outside of the turn, is to apply a 'countersteering pressure' on the handlebars. You push the bars to the outside of the curve. This is intuitive, and most of us aren't even aware of it.

Leaning the Bike

When you ride no-hands, you still can go around curves. How do you countersteer without your hands on the handlebars? Let's assume you make a right turn. You lean your body to the right, toward the inside of the curve. Many believe that the bike now will lean to the right, but it's not that simple: Inertia keeps the bike's center of gravity in the same location. To compensate for your body leaning to the right, the bike actually leans to the left – to the outside of the curve.

By leaning the bike to the left, you invoke the forces of balancing: Gyroscopic forces and the weight of the handlebars steer the front wheel to the left. As a result, both wheels move to the left – to the outside of the curve – just like they did when you actively countersteered on the handlebars. The result is the same: The bike now starts to lean to the right.

Leaning your body (right) does not directly lean your bike into a turn. Indirectly, it turns your handlebars into a countersteer, which causes your bike to lean.

You can steer the bike by leaning your body even when you aren't riding no-hands. Some call this 'steering from your hips.' Countersteering through leaning the bike, as opposed to countersteering it directly, is a relatively slow process, because you use the gyroscopic forces and the weight of the handlebars to countersteer the front wheel. Riders who steer from the hips probably also instinctively move the handlebars.

In short, to initiate a right turn, you can either turn the handlebars to the left or lean to the right. Both actions will move your front wheel to the left, unbalance the bike, and initiate a lean to the right.

Steering Into the Turn

As long as you apply countersteering pressure on the handlebars, the bike continues to increase its lean angle. Once you stop applying pressure on the handlebars, the gyroscopic forces of the front wheel automatically steer the bike into the turn and stabilize its lean. Now gravity and centrifugal force are balanced. Without any input, a good bike is self-stable while cornering – it will continue on its line through a corner.

As you corner, your handlebars are turned slightly in the direction of the turn. The more you lean, the less you need to turn the handlebars. Thus, the faster you go, the less you turn the handlebars.

Straightening the Bike

When you are done with your (right) turn, you need to bring the bike back to its upright position, so you can go straight again. Since a bike is balanced as it corners, just like it was on the straight, the process is the same as when you initiate a left turn. You apply countersteering pressure to the right and move the wheels further into the turn. The bike now rotates out of the lean, aided by centrifugal force.

Once the bike is upright, you stop applying pressure on the handlebars. Now you are going straight again.

Handling Summary

Riding a bike is simple, and the inputs to balance a bike on the straight are the same as they are to balance a bike during cornering: You countersteer to move the wheels sideways until the bike is balanced at the lean angle that corresponds to the radius of the corner and your speed. Simple pressure on the handlebars is all you need to initiate a lean and to right the bike again. Once you remove the pressure, the bike continues on the selected course. This self-stability makes bikes easy and intuitive to ride.

Countersteering is how we walk and run around corners, so it's completely intuitive. A good bike's geometry is designed to work with inputs that are intuitive for its rider.

Geometry

"A good bike goes where the rider is looking."

A bike's geometry translates the rider's small, intuitive inputs into directional changes. Many riders are unaware of the fact that they countersteer to get the bicycle to go where they want, because the inputs mimic those of walking and running. The best-handling bicycles are those that readily accept these very small steering inputs.

If the bike initially resists the input, the rider has to increase the force on the handlebars. Most likely, the resulting input will be too large. The bike will 'overshoot' and a correction is required. As a result, the bike requires constant corrections, which makes its handling less precise.

If the bike over-reacts to very small inputs, it will also require constant corrections to hold its line, especially under a tired rider. How a bicycle reacts to steering inputs is determined by a number of factors.

Gyroscopic Force

The gyroscopic forces of the spinning wheels don't keep the bike upright, but they stabilize it in several ways. Gyroscopic forces increase with speed: Your bike becomes more stable as you go faster. At very high speeds, gyroscopic forces predominate, and all other factors become less important. That is why most bikes handle similarly at very high speeds.

The spinning wheels generate gyroscopic forces that make the bike harder to lean. The gyroscopic force of the front wheel also exerts a self-centering force: When you turn the handlebars, the gyroscopic force tries to return the front wheel back to a straight-ahead position. Both of these factors make bikes with heavier and/or larger wheels more reluctant to turn.

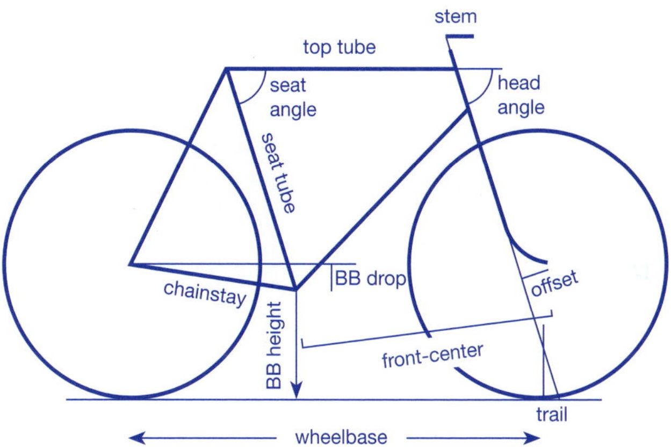

A bike's geometry is the result of the interaction of the lengths and angles of many parts.

Geometric Trail

A bike's steering axis intersects the ground ahead of the tire's contact patch. The distance between these two points is called 'geometric trail' or simply 'trail' (see drawing on p. 79).

When the front wheel turns, the contact patch moves sideways – imagine the caster of a shopping cart. Geometric trail exerts a force on the wheel that straightens it. (In reality, it's not quite so simple, because the steering axis is inclined and the bike is leaning as soon as the front wheel is turned, which introduces a host of complex interactions.)[24]

Geometric trail is often used to characterize a bike's handling. In the past, many believed that more trail made a bike stable, and less trail made it 'twitchy.' We know today that this isn't necessarily the case. Geometric trail tends to stabilize the bike, but there are other effects related to trail that can counteract this.

Wheel Flop

One of the most important factors determining a bike's handling is wheel flop. The front wheel is offset from the (inclined) steering axis. As the fork turns, the front of the bike lowers. If you want to visualize this, clamp your bike in a repair stand with the wheels barely touching the ground. Turn the fork, and the front wheel will lift off the ground.

Wheel flop lowers the front of the bike when the handlebars are turned.

Any weight on the front of the bike's frame will push the front wheel further to the side to which it is already pointing. Wheel flop acts like a car's power steering by amplifying the rider's initial steering inputs.

A bicycle's fork never points straight ahead for long, since the rider is balancing the bike by countersteering. Imagine a ball balanced on top of a knife's edge. It's impossible to center the ball perfectly – it will always roll down one side or the other.

Wheel flop tends to turn the handlebars, but other forces, like geometric trail, stabilize the bike. Playing these forces against each other results in a bike with pleasant handling.

Wheel flop is determined by the combination of head angle and geometric trail. A shallower head angle increases wheel flop, and so does more geometric trail. You can see this in the formula for the wheel flop factor *f*, which expresses how much the front of the bike lowers as the handlebars are turned:

$$f = \text{trail} \times \sin(\text{head angle}) \times \cos(\text{head angle})$$

The second factor that determines a bike's wheel flop is how much weight rests on the headset. A racer with a low position puts more weight on the front of the bike, so even with the same geometry, they will experience more wheel flop than a city rider with an upright position. In practice, builders compensate for this by reducing the wheel flop factor on racing bikes – usually via a steeper head angle.

Trail vs. Wheel Flop

Trail makes the bike more stable, but it also increases the wheel flop factor. At first sight, it seems like the two factors cancel each other. Reality is more complex, because many of the forces involved are dependent on speed.[25, 26]

At low speeds, the influence of wheel flop predominates. Bikes with too little trail (and little wheel flop) are hard to steer – like a car with no power steering. They tend to go straight, and riders hit potholes that they thought they could avoid. At higher speeds, bikes with too little trail are stable on the straights and corner with ease, but they tend to drift off their line in mid-corner.

Bikes with too much trail (and too much wheel flop) tend to veer off course at low speeds. Going uphill, their riders may find themselves unintentionally on the edge of the road. Or they touch handlebars with the rider next to them.

At higher speeds on straight roads, bikes with too much trail (and too much wheel flop) resist turning the handlebars at first (when the handlebars are still pointed straight-ahead), requiring more force to balance the bike, but once the handlebars are turned, the wheel flop turns the bars further, and the bike tends to veer off course, requiring a correction.

Cornering, a high-trail bike first resists turning, but then dives eagerly into the turn as the wheel flop amplifies the initial handlebar movement. Once into the turn, the wheel flop steers the front wheel further into the curve. This pulls the bike upright again. Bikes with a lot of wheel flop have a very prescriptive way of cornering: Initiate a turn, fall into the turn, pull the bike upright, go straight again. This works fine for the 90° corners of a criterium course, but these bikes can have trouble rounding long hairpins in the mountains where the same lean angle must be maintained for longer periods of time.

The handlebars of bikes with optimal levels of geometric trail (and wheel flop) are easy to turn, but they don't overshoot the rider's initial inputs. Unless the handlebars are turned further, the bike will not change its course. Because these bikes react in a linear fashion to steering inputs, they can be placed on the road with great precision. However, they require a light touch on the handlebars, as they react to small inputs. For riders who tense up on the handlebars, more trail may be required to resist small, involuntary handlebar inputs.

Bikes with little wheel flop require setting up for a turn. That is how you ride at medium and high speeds, but at low speeds in technical terrain, more wheel flop is required to change direction quickly. That is why mountain bikes have geometries with a lot of wheel flop.

Head Angle

Even though the wheel flop factor and trail are directly related, it is possible to fine-tune each variable independently by adjusting the head angle. Bikes with steeper head angles have a smaller wheel flop factor for a given geometric trail measurement.

This is useful when adjusting the geometry for the weight distribution of rider and bike. With a more upright riding position or a load on the rear, the front of the bike is less loaded, and a greater wheel flop factor is needed to provide just the right amount of power steering. This is achieved via a slacker head angle. That is why many traditional touring bikes – designed for a more upright position and rear loads – have slacker head angles than racing bikes.

Wheel flop and geometric trail are not the only factors that influence bicycle handling. Among the other factors that determine how a bike steers, two are especially important: pneumatic trail and any load that is attached to the fork.

Pneumatic Trail

The side forces on the contact patch of the front tire stabilize the bike in a similar way to geometric trail. The tire's contact patch deforms when the bike steers: At the front, the tire's rubber is laid down on the road without lateral forces acting on it. (Before the tire touches the ground, there are no sideforces.) When the bike corners, the bike is pushed

During a turn, deformation of the tire contact pulls the front wheel straight and stabilizes the bike.

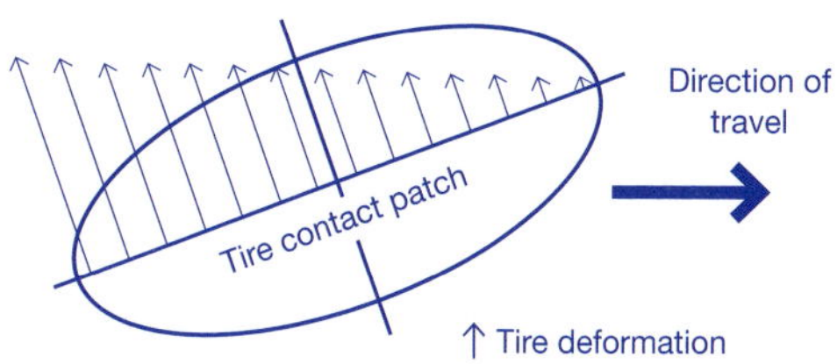

toward the outside of the turn, but friction holds the tire on the road surface. The tire deforms more and more while the contact patch touches the ground: The deformation increases toward the back of the contact patch. As a result, there is more sideforce in the rear half of the contact patch than in the front half. This tends to center the bicycle's fork toward the straight-ahead position. Once the rubber leaves the road, it springs back into its initial shape.

Narrower tires have a longer contact patch than wider tires, and thus *should* have more pneumatic trail. In practice, the opposite is the case: Wider tires stabilize a bike's steering more than narrow tires. There are a number of reasons for this: Wider tires run at lower pressures, so their contact patches are larger. Furthermore, their taller sidewalls can deform more in response to the forces of pneumatic trail. It appears that hard, narrow tires slip more at the contact patch, which resets the forces as pneumatic trail builds up: Their self-centering force never gets as large as it does with wider tires.

Pneumatic trail adds stability, so bikes with wider tires should be designed with less geometric trail. With the same geometric trail as narrow-tired bikes, bikes with wider tires tend to feel sluggish in their steering. Adjusting one's line in mid-corner can be difficult on these bikes. On the other hand, using narrow tires on a bike optimized for wide tires can result in too little directional stability.

Knobby tires have less pneumatic trail, because they don't have a continuous contact patch on the road surface.

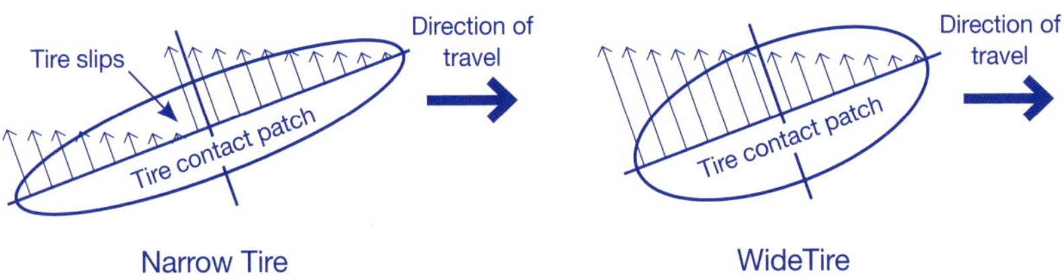

Narrow tires (left) have less pneumatic trail than wide tires (right). Apparently, a narrow tire is too stiff to deform much. It slips as the deformation builds up.

Pneumatic trail is also reduced on unpaved surfaces, where tire slip reduces the side forces generated by the front tire. Bikes that are ridden primarily on gravel sometimes have slightly more geometric trail to compensate for the reduced pneumatic trail.

Load

Luggage on the bike, whether panniers, under-seat or handlebar bags, can have a significant effect on a bike's handling.[27] On page 215, we discuss why front loads are easier to balance than rear loads. Here we will look at how to design a bike for a front load. Front loads are usually attached to the fork, so they influence the steering in two ways:

Tendency to Turn the Fork

Because the steering axis is inclined, any weight in front of the steering axis has a tendency to turn the fork. (This is often confused with wheel flop, but it's a different force.) This is especially noticeable when carrying a load on a front rack. The tendency of a fork-mounted load to turn the wheel is most pronounced at low speeds, because many factors stabilizing the bike (trail, gyroscopic forces) increase with speed and are not very effective at low speeds.

The tendency of a load to turn the fork is greater with shallower head angles.

The tendency to turn the fork is greater with slacker head angles. On a bike with a 90° head angle, the load would not turn the fork. On the move, the load's tendency to turn the fork amplifies the effects of wheel flop. This is especially pronounced when riding uphill: Speeds are lower and the head angle decreases as the bike is tilted upward.

Inertia Stabilizes Steering

Once the bicycle is moving, a front load stabilizes the steering by providing inertia that resists turning the fork. This feels similar to a steering damper.

Both effects – the tendency to turn the fork and the inertia that stabilizes the steering – increase with heavier loads and with greater distance of the load's center of gravity from the steering axis. One reinforces wheel flop, the other stabilizes the bike, so they counteract each other. The magnitudes of each effect depends on the front-end geometry of the bike.

On bikes with shallow head angles and lots of wheel flop, the tendency of the load to turn the fork tends to dominate, thereby reducing the bike's directional stability – hence the bad reputation of handlebar bags when used with high-trail geometries of racing or rear-loading touring bikes.

On bikes with steep head angles, the inertia of the load is more important, so a load makes the bike more stable. That is why bikes with low-trail geometries and relatively little wheel flop work especially well with handlebar bags.

The inertia of a fork-mounted load can stabilize the steering.

Frame-Mounted Racks

Front racks can be attached to the frame rather than the fork. This is a good idea for delivery bikes, because it eliminates the load's tendency to turn the fork and improves the handling at low speeds. But at higher speeds, a frame-mounted rack has no inertia to stabilize the steering.

Delivery bikes tend to wander or even shimmy when ridden at high speeds – but they aren't intended for that. In France, postal carriers – who stop at every house – rode bikes with frame-mounted racks. The famous newspaper couriers, the *porteurs*, who rode at high speed between newsstands, mounted their racks on the fork.

Frame-mounted racks are more stable at low speeds, but less stable at high speeds.

Wheel Size

Gyroscopic forces exert a self-centering force on the steering, which adds to the bike's stability. Gyroscopic stability is influenced by the wheel's mass (m) and its size (r = radius). The moment of inertia (I) is calculated as follows:

$$I = m \times r$$

A larger wheel spins slower at the same speed, which cancels the effect of its larger radius, so only the weight of the wheel (and how that weight is distributed) matters. Larger rims and tires use more material, so larger wheels inevitably are heavier.

Heavier wheels have a greater moment of inertia, and thus make the bicycle more stable. Bikes with wheels that are too heavy will be too stable: It is difficult to initiate a turn or adjust the bike's line in mid-corner.

If the wheels are too light, the bike will lack stability on the straights and tend to wander off-course if the rider does not pay close attention. In corners, the bike will not hold its line. Its trajectory is less predictable.

Bicycle Quarterly tested three bikes with identical geometries, but different wheel sizes: 700C, 650B and 26". Three riders with different preferences – a racer, a commuter, a randonneur – rode the bikes with tires of different sizes and weights.[28] When lightweight tires were mounted, all riders preferred the 700C bike with narrow tires; the 650B bike with wide tires; and the 26" bike with ultra-wide tires. When the tires were replaced with heavier ones, each rider noticed the increase in stability and preferred the next-smaller wheel size. Based on this experiment, we calculated the rotational inertia for optimum handling at 0.30 – 0.35 kg m.[29]

Larger and/or heavier wheels make a bike more stable.

In practical terms, this means that wheel size should scale with tire (and rim) weight. To offer the same handling, heavier tires and rims should be used with smaller wheel sizes. And since wider tires are heavier, they benefit from smaller wheels, too.

Even though our three testers all preferred the same tire/wheel combinations, some riders might prefer more or less rotational inertia. It appears that only very few riders prefer bikes that are less stable than what our testers found optimal.

Some riders prefer bikes that are more stable. This feeling is often described as 'cornering on rails': Once the rider has selected a line, the bike will remain on this radius, even if the rider tenses up on the handlebars. This gives a feeling of stability, where the bike will continue on its path, ignoring the rider's inputs. The same stability makes it difficult to adjust

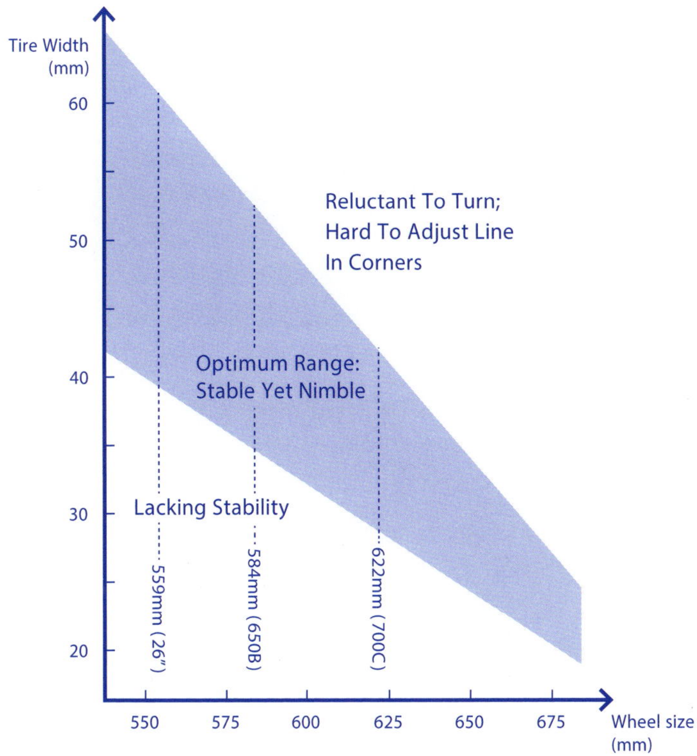

Different tire widths require different wheel sizes for optimum handling: Upper part of shaded area is for light tires and rims, lower part for heavier tires/rims.

the line in decreasing-radius corners or to avoid unforeseen potholes or obstacles on the road.

For most riders, the optimum wheel sizes for lightweight high-performance tires on road and all-road bikes are:

Aluminum rims:
27 – 35 mm tires: 700C (622 mm)
35 – 45 mm tires: 650B (584 mm)
40 – 55 mm tires: 26" (559 mm)

Carbon rims:
30 – 44 mm tires: 700C (622 mm)
45 – 53 mm tires: 650B (584 mm)

There are no lightweight 26" carbon rims at this time. For aluminum rims, there is some overlap between 650B and 26" wheels, because the size difference is small. For tires narrower than 27 mm, the wheels ideally would be bigger than 700C, but larger wheels are not currently available. For tires wider than 55 mm, wheels smaller than 26" may be advisable.

Bikes with wheels that have gyroscopic inertia outside the 'ideal' range can ride very nicely, so don't worry if your bike falls outside this range. But when the time comes to replace your bike, choosing a different wheel/tire combination can improve the handling of your new bike.

Why not adjust the front-end geometry to compensate for rotational inertia outside the 'optimum' range? Unfortunately,

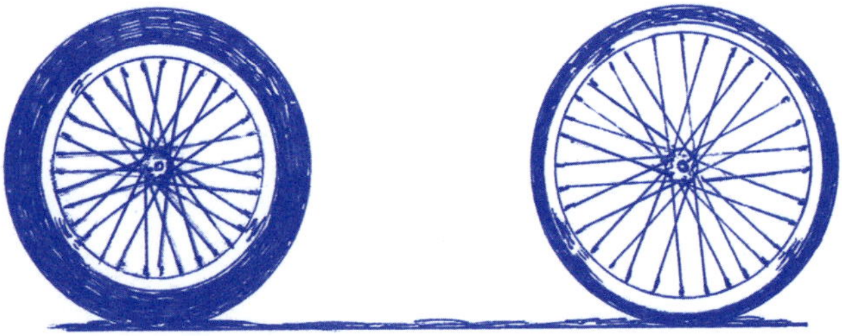

Larger tires can be paired with smaller rims to keep the rotational inertia the same as on narrow-tire performance bikes. The overall diameter of the wheel should be slightly smaller to account for the heavier weight of the wider tire.

How geometries evolved

During the 1960s, narrow tires became popular. To compensate for their reduced pneumatic trail, most builders reduced the fork offset to increase trail. Some also steepened the head angle to prevent wheel flop from becoming too pronounced. That is the geometry most racing bikes still use today. When ultra-narrow tires (18-20 mm) were popular during the 1980s and 1990s, these geometries were the only way to make bikes with sufficient stability.

As tires have become wider again in recent years, the added pneumatic trail does not work well with the relatively high geometric trail of most racing bikes. It's likely that racing and all-road bike geometries will evolve toward less geometric trail in the future to improve these bikes' handling.

For much of the 20th century, tires on bicycles became narrower, before getting wider again in recent decades.

this doesn't work: If the trail is increased to make the bike more stable, wheel flop also increases, which in turn reduces the stability. The two factors roughly cancel each other, which is why wheel size is so important for dialing in the bike's handling.

Small-wheeled bicycles are inevitably less stable, unless you increase the weight of the wheels to obtain the same rotational inertia as an 'optimally-sized' wheel. Alex Moulton machined a steel insert for the rims of his iconic small-wheeled bicycles to increase the rotational inertia and make up for this deficit. A better solution is to increase the tire width. The wider tire is heavier, increasing the rotational inertia, and its greater pneumatic trial also increases the bike's stability without increasing wheel flop.

Larger Wheels for Larger Bikes?

Should a bike's wheel size scale with the frame size? Should a taller cyclist ride on bigger wheels? Only if the taller rider wants a bike that is less nimble and more stable. For smaller riders, smaller wheels may be necessary to design bikes that don't have toe overlap, even if this makes the bike a bit less stable than would be ideal otherwise.

Larger riders should choose their wheel size to obtain the desired handling, not to scale the wheels to their frame size.

Other Factors

The five factors described so far – geometric trail, wheel flop, pneumatic trail, load and wheel size – explain most of the handling traits we have observed in a large variety of bicycles. These factors interact with each other. If they are well-balanced, very different bicycles can have similar – and excellent – handling. Other factors are sometimes thought to influence a bike's handling. Let's examine them in detail.

Bottom Bracket Height

A lower bottom bracket lowers a bike's center of gravity. Does that make the bike handle more nimbly? This effect is very small: The center of gravity of an average-height cyclist is about 900 mm (35") above the ground, yet bottom bracket height varies only by up to 20 mm (0.8"). The resulting change in the center of gravity is just 2-3% – too small to make a significant difference.

However, bottom bracket height is important when you put a foot down during a stop. The standard bottom bracket height for road bikes is 265 mm, which just allows putting a foot down while seated on the saddle.

Bottom bracket height also affects the cornering clearance of the bike. Some criterium bikes have high bottom

The range of bottom bracket heights on most bicycles is too small to affect the center of gravity.

brackets, so riders can pedal through corners. Professional racers tend to arrive at corners too fast to continue pedaling: They lean so hard that even a high bottom bracket cannot prevent pedal scrape. Greg LeMond is reputed to have said: “If you are pedaling through a corner, you weren’t going fast enough on the straight.”

Chainstay Length and Wheelbase

Bikes with longer wheelbases are sometimes thought to be more stable. And yet, the wheelbases of most performance bikes fall within a narrow range between 995 and 1045 mm. The difference of 50 mm is just 5% of the total wheelbase – too little to make a significant difference.

An experiment confirmed this: Peter Weigle once built a bike with the same geometry as a 1954 Alex Singer, except he lengthened the chainstays by 25 mm to fit a pump behind the seat tube. Peter and I went for a ride on these two bikes and swapped bikes several times. Even riding them back-to-back, we found that their handling and comfort were indistinguishable. (Both bikes were equipped with identical wheels and tires.)

Even a large difference in chainstay length represents only 5% of the total wheelbase: too little to affect a bike’s stability.

Chainstay length may affect the comfort of a bike. Longer chainstays place the rear axle further back, and thus further away from the rider. Since the rider on a typical bike sits between 200 and 230 mm ahead of the rear axle, increasing the chainstay length by 25 mm will increase this distance by 10%. It is unclear whether this is an important factor, as tire size and pressure seem to have a bigger impact on comfort.

Chainstay length is mostly determined by the sizes of wheel, tire and fender. Shorter chainstays are stiffer, which may help tune the 'balance of frame stiffness' (see p. 64) for optimum performance. That is why many builders try to keep the chainstays short. (The weight savings are minimal.)

Front-Center

Front-center is the distance between the center of the crank spindle and the center of the front wheel. Front-center affects the weight distribution of the bike. However, the differences in front-center for a given frame size are relatively small. Compared to the large differences in weight distribution between tall and short riders (see p. 124), small differences in front-center don't appear to be important.

Front-center does affect toe overlap. For a description about how to adjust front-center, see the chapter on frame design (p. 121).

Front-center determines whether a bike has toe overlap.

High-Speed Stability

Most bikes, no matter how much or how little trail they have, are stable at high speed. During Bicycle Quarterly's tests of front-end geometry, we rode a mid-century bike with just 11 mm geometric trail: At very high speeds, this bike's handling was indistinguishable from bikes with much more trail. A bike's high-speed stability appears to be derived mostly from the gyroscopic self-centering forces of the front wheel and other factors unrelated to trail.

Shimmy can make high-speed descending scary and even dangerous (see p. 228). Crosswinds can also be a problem at high speeds: The bike travels so fast that it gets thrown off course before the rider can correct.

Most bicycles are stable at high speeds.

Crosswinds

When a crosswind gust hits the rider, the bike will lean downwind and thus begin to turn into that direction. This is the same mechanism that makes the bike turn when riding no-hands. The change of direction caused by the leaning of the bike alone happens slowly, and the rider instinctively corrects for it.

When bikes are thrown off course, it's caused by a slightly different mechanism: As the wind pushes the bike sideways, the bike's geometric trail provides a lever arm for the ground force to move the steering.[30] This means that the sideways force of the wind will turn the fork, countersteering the bike toward the wind. This reinforces the leaning of the bike caused by the wind pushing on bike and rider.

The more trail the bike has, the longer is the lever that turns the fork. On high-trail bikes, sudden gusts of wind cause rapid changes of direction that can be difficult to counter. Low-trail bikes, on the other hand, give the wind much less leverage over the steering. The entire bike will move sideways, but it will not suddenly change its direction. As a result, low-trail bikes are affected less by crosswinds.

Trail (the distance between steering axis and wheel contact) provides a lever for crosswind forces to turn the fork.

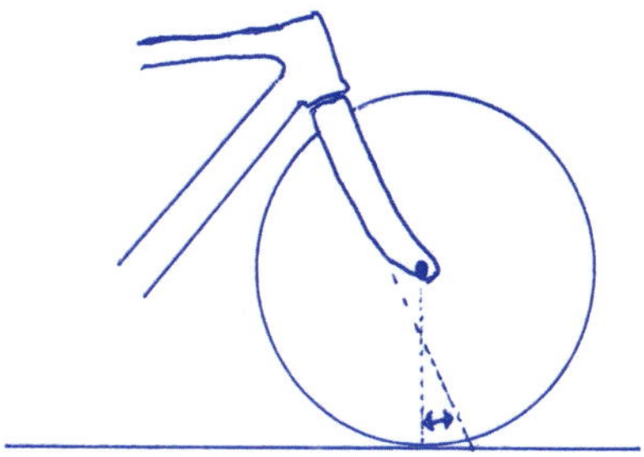

Sample Geometries

In the previous chapter, we have seen that numerous factors affect a bike's handling. Most riders understand that replacing the fork with one that has different offset (rake) can change a bicycle's handling, but not everybody knows that a different tire size or load placement can have similar effects.

For Bicycle Quarterly, we've tested and measured more than 100 bikes. They range from carbon racing bikes to porteurs designed to carry 50 kg (110 lbs) of newspapers on a front rack. There have been city bikes, camping bikes, cyclocross bikes, gravel bikes and tandems – bikes for every purpose (except track racing). We've ridden each for several weeks and at least 200 km (125 miles). We've also measured many classic bikes from a time before racing bikes got all the attention; a time when all-road bikes for gravel roads and camping adventures were highly sophisticated machines, rather than relegated to the back pages of bike makers' catalogs. To confirm our findings, we've worked with bike builders (Jeff Lyon, Kogswell, Peter Weigle) to build multiple bikes with specific geometries. These bikes have allowed us to test how different parameters affect a bike's handling.

We've found that it is possible to build very different bikes with the same intuitive, excellent handling: Stable on the straights, nimble in corners, and relatively easy to ride no-hands (with the exception of the porteurs that carry too heavy a load for no-hands riding). This is achieved by adjusting the geometry for all factors that matter: riding position, tire width, any load the bike carries, even the speed at which the bike will be ridden.

Most geometries showcased in this chapter would be considered 'low trail.' These bikes require a light touch on the handlebars. Like a good horse, they must be guided rather than forced on the rider's chosen course. Riders who tense up and grip the handlebars tightly may not find these geometries to their liking: The precise steering reacts to every handlebar movement, leading to involuntary steering inputs if the rider is tense. For riders unable to relax their grip on the bars, bikes with more trail may be preferable, even though their handling tends to be less precise.

The sample geometries shown in this chapter are taken from actual bikes. Their frame sizes have been adjusted to make them easier to compare.

Racing (Road) Bikes

The industry likes to call racing bikes with narrow tires 'road bikes' to increase their appeal to non-racers. For most road riding, all-road bikes with wide tires are a better choice.

Racing bikes with narrow tires tend to handle similarly even with very different front-end geometries. The forces of geometric trail (more stability) and wheel flop (less stability) counter each other. Racing bikes with more trail also have more wheel flop and so their overall stability is similar to those with less trail and less wheel flop.

However, as racing bikes adopt wider tires, the added pneumatic trail changes the equation. To offer the same intuitive handling, bikes with wider tires should use less geometric trail. The added advantage is that they have less wheel flop, which helps when climbing at lower speeds, when riding over bumpy terrain, and in cross-winds.

Designed for 28 mm tires, this racing bike has a 73° head angle and about 49 mm of trail. Most racing bikes today have more trail, because their geometries were developed for narrower tires.

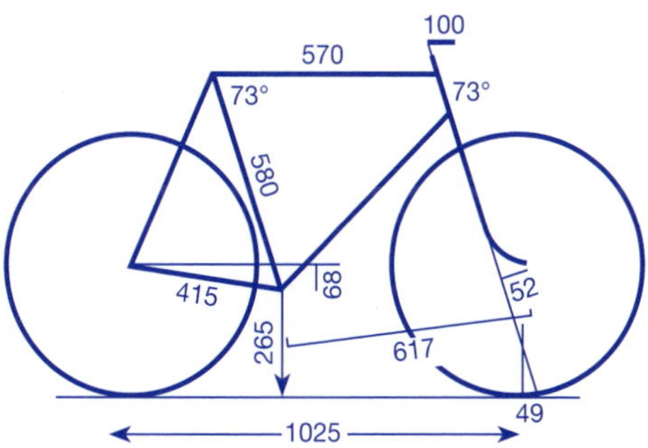

Gravel and Cyclocross

Cyclocross and especially gravel courses usually are flowing and fast, rather than technical and tight, so the ability to handle predictably at high speeds is important – and that requires a road bike geometry with adjustments for wider tires.

The Alan was the classic cyclocross bike of the 1970s and 1980s. From 1976 until 1990, twelve professional world championships were won on Alans. The geometry of this 1980s Alan Cyclocross bike is similar to that of the racing bike on the previous page. The head angle is a bit steeper to reduce wheel flop, and the bottom bracket is higher to allow pedaling with upside-down pedals/toeclips. With low-pressure tires, the resulting bike is very stable on rough terrain, even at low speeds. Climbing steep hills over rocks and roots is relatively easy, as the bike goes where it is pointed. During high-speed descents, the bike remains nimble, making it easy to steer around rocks and other obstacles. For all-road and gravel bikes with wider tires, increase the fork offset by 5 mm (42 mm tires) to 10 mm (52 mm tires). This reduces the geometric trail to compensate for the wide tires' greater pneumatic trail.

This geometry does not work well on some very tight and technical 'cross courses, such as when a large field is taped-off into a labyrinth with multiple tight 180° turns. For these ultra-tight turns, this geometry does not have enough wheel flop to make quick turns of the handlebars intuitive. The bike tends to run wide. In these situations, a monstercross geometry (see next page) works better.

Alan cyclocross bike for 28- to 33 mm-wide tires: The bottom bracket can be lower, since today's riders no longer use toeclips.

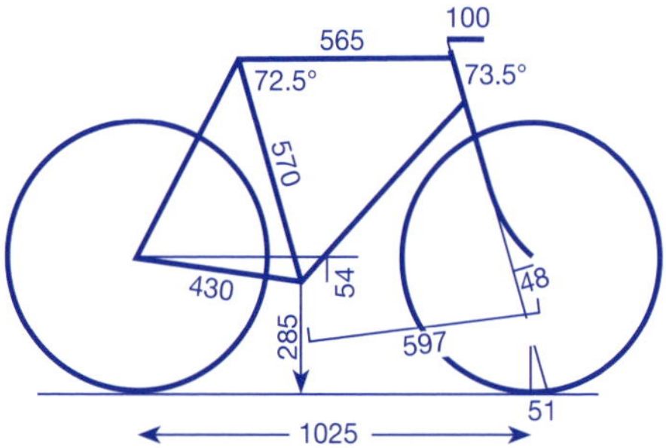

Monstercross/Singletrack

Singletrack requires a different geometry with more wheel flop. On singletrack, low speeds and tight corners require turning the handlebars quickly and far. More wheel flop is needed to provide 'power steering,' so the bike reacts to intuitive inputs under these conditions. To compensate for the wheel flop and make the bike stable at high speeds, more trail is needed as well.

To achieve this, monstercross geometries are inspired by mountain bikes, with shallow head angles (more wheel flop, more trail) and moderate fork offsets. A more upright position and wider handlebars are useful to increase the steering leverage for extra 'power steering.' See page 235 for more about the difference between all-road and monstercross bikes.

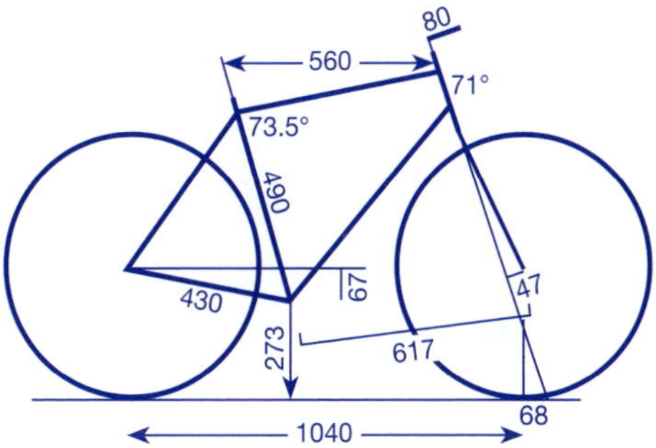

Monstercross geometry has more wheel flop for quick turns in tight singletrack.

Randonneuse

Randonneur bikes are designed for riding long distances at high speeds, unsupported, in all weather conditions. They carry a moderate load in a handlebar bag, where it is easily accessible while riding.

The handlebar bag's inertia stabilizes the bike, so its geometric trail is reduced to 39 mm, compared to the racing bike's 49 mm. This geometry relies on the pneumatic trail of medium-width tires for its stability. With 30 mm tires, it is close to perfect, whereas with narrow tires at high pressures, it lacks directional stability. We have not yet found a way to create a bike with narrow tires that can carry a front load and offer good handling. Fortunately, few riders use narrow tires for long-distance cycling any longer, so this has become a moot point.

Wider tires add pneumatic trail, so a randonneur bike with 650B x 42 mm tires benefits from 5-10 mm more fork offset than the 700C bike (and correspondingly less trail). The 650B bike handles very similarly to the 700C randonneur bike, but with increased low-speed stability due to its decreased wheel flop factor. Despite their low geometric trail, these bikes are very stable at high speeds of 80 km/h (50 mph) and above.

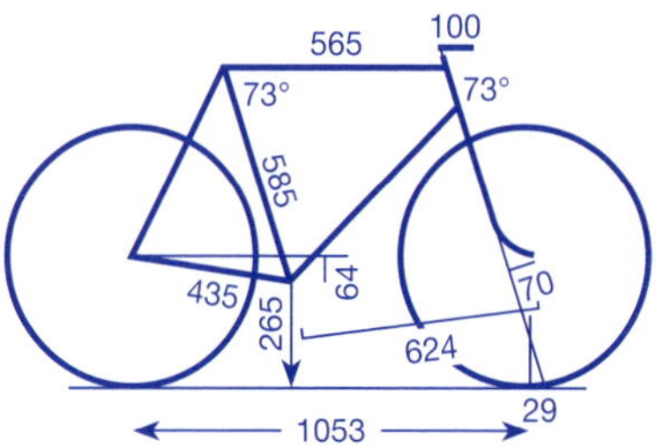

650B

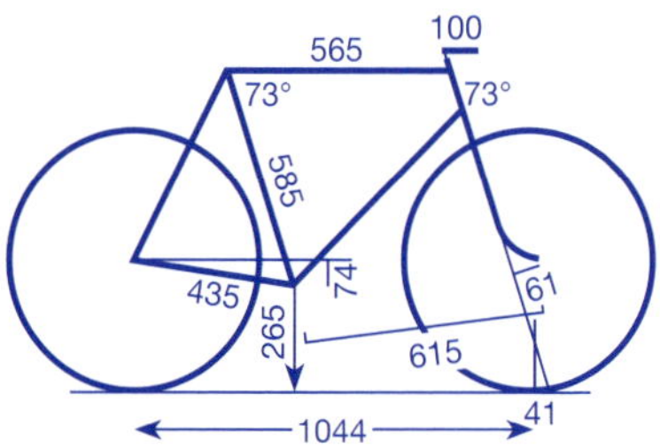

700C

Randonneur bikes use handlebar bags that stabilize the steering, so they have less trail than unloaded bikes. Medium-width 700C tires require more trail than wide 650B tires.

Porteur and Camping Bikes

A heavy load (>5 kg/11lb), whether it's carried on a porteur rack or a front low-rider, greatly increases the inertia on the steering. So much in fact, that more wheel flop ('power steering') is needed, so the rider can guide the bike with a light touch on the handlebars. We've found that a trail figure of 40 mm works well both with heavily-loaded low-rider racks and with a porteur rack. When riding these bikes unloaded, they will be slightly more sluggish in their response to steering inputs, but the difference is small, and they still handle pleasantly.

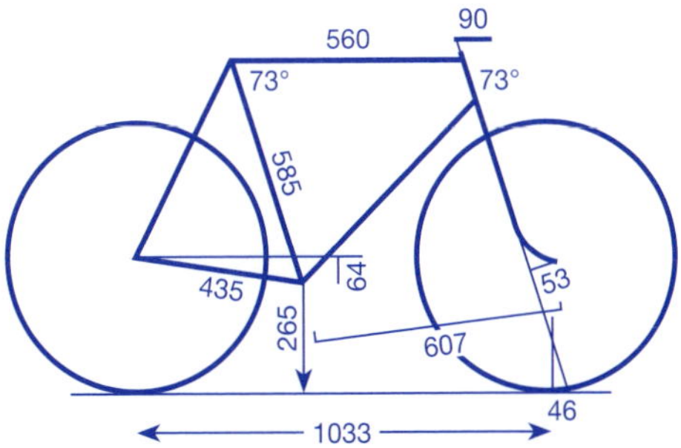

Bikes for loaded touring have more wheel flop to overcome the inertia of a heavy front load.

City Bike and Rear-Loading Tourer

A city bike is designed for a more upright position and a more rearward weight distribution. To compensate for the reduced load on the front wheel, the wheel flop factor has been increased by slackening the head angle to 71.5°, similar to a traditional rear-loading touring bike. With its wide 650B x 42 mm tires, this bike is very stable, even at low speeds. The geometry for bikes with a rear load is similar, but rear loads generally make a bike handle less predictably than front loads (see p. 215).

Bikes for upright positions and rear loads have slacker head angles.

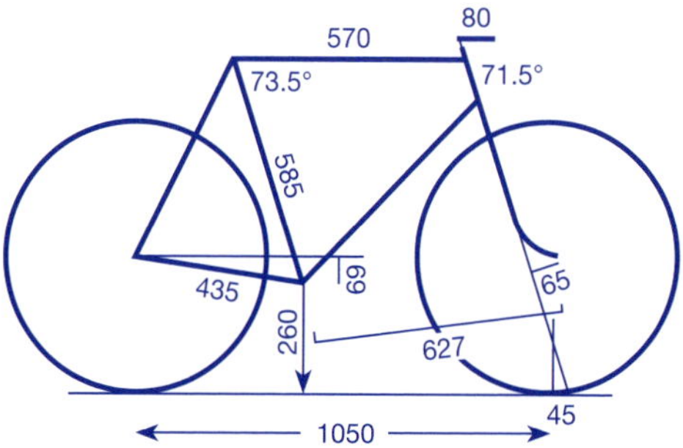

Tandem

Tandems carry the entire weight of the front rider (captain) on the front of the bike, whereas a single bike carries about 35% on the front. This means that a tandem has roughly three times as much wheel flop as a single bike.

The geometry of a tandem should be adjusted to account for the increased wheel flop, otherwise, the tandem will tend to veer off course every time the rear rider (stoker) moves. Good tandems have steep head angles and long fork offsets, which works out to low trail and a low wheel flop factor.

The tandem shown here handles very much like the single bikes in this chapter. Unlike tandems with more wheel flop, this tandem is easy to ride with inexperienced stokers, because it is not affected by small stoker movements. In fact, it can even be ridden no-hands, which is something few tandems allow. Switching between this tandem and the single-bikes shown in this chapter is entirely intuitive and does not require re-acquaintance, because all these bikes react similarly to steering inputs. Like the randonneur bikes on p. 102, the tandem relies on the pneumatic trail of its wide tires for stability.

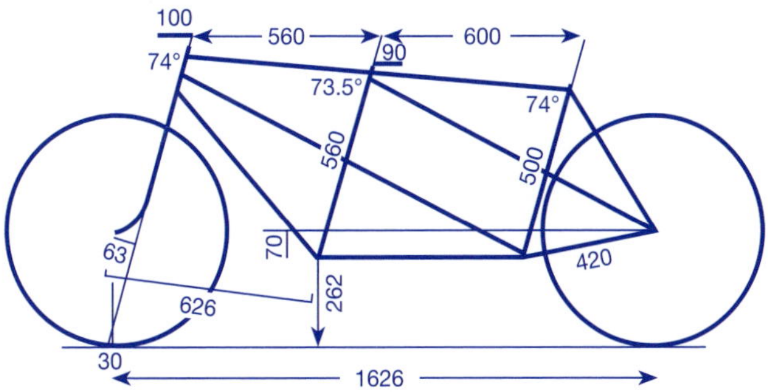

Tandems have steep head angles to reduce the effect of the captain's weight on the steering. Wide tires (650B x 42 in this case) are essential for stability.

Similar Handling for All

All these bikes offer the same precise handling, because their geometries have been adjusted for their weight distribution, tire size and other factors. There are many other geometries that work reasonably well. It is almost impossible to build an unrideable bicycle, and even bicycles with less than optimal geometries can be great fun to ride.

There is no need to overthink this, either. The geometries shown in this chapter can be adapted by builders who understand the underlying forces. Bike geometry can appear complex at first, but many factors have only relatively small effects on how a bike rides. Focusing on the important parameters – geometric trial, wheel flop, pneumatic trail and wheel size – enables riders to assess a bicycle's geometry and determine whether it is likely to suit their needs. A test ride is useful to confirm whether a bike will work for you.

Bikes as different as racing and camping bikes can have similar handling, if the geometry is adjusted for weight distribution, tire size and other factors.

Mid-Trail Geometries

It may be tempting to use a mid-trail geometry (typically with 55-60 mm geometric trail) as a compromise between low and high trail. In the real world, this doesn't always work well. In high speed corners, mid-trail bikes have too much trail to offer precise and easily adjustable handling. At moderate speeds, they have too much wheel flop to be stable, especially on bikes with narrow tires and little pneumatic trail. Conversely, at low speeds, they have too little wheel flop to offer the maneuverability of high-trail bikes. The factors of trail and wheel flop are not linear. Real-road experience has shown that high-trail and low-trail bikes feel more similar and intuitive in their handling than mid-trail bikes.

The one place where mid-trail geometries work well is criterium racing bikes. These bikes are designed to handle well in 90° corners taken at high speeds. Ample wheel flop – and the low pneumatic trail and low gyroscopic stability of narrow tires – make the bike fall into the lean as soon as the rider initiates a turn. The rider corrects to 'catch' the bike. The wheel flop reinforces this move and rights the bike as the rider launches into the next straight. Cornering is done in a quick sequence of falling into the turn and righting the bike. These bikes struggle to corner on a constant radius, for example, in long hairpin turns during mountain descents. But that is not what they are designed for...

Comfort

"Even after riding for more than 40 hours, I am flying. I am glad to have changed my handlebars. My hands, wrists, arms, shoulders and neck are free of pain or even discomfort."

In the earlier chapters about how to make your bike faster we already touched on the subject of comfort, because comfort and speed are so inextricably linked. Comfort on a bike is complex because it depends on the interaction between rider and machine. Let's start with the simplest part, the absorption of vibration and bumps.

Unsprung Mass

A good suspension system deals with vibration as close to the source as possible. The goal is to keep the unsprung mass – that part that moves up and down – as small as possible. This applies to bicycles as well.

At a speed of 25 km/h (15.6 mph), the vibration of rough pavement (aggregate size: 7 mm/0.25") occurs at about 500 Hz. This means that pavement irregularities hit the bicycle's wheels 500 times per second. Even on coarse cobblestones like those found in Paris-Roubaix, the frequency is still about 40 Hz.

Alex Moulton's tests during the development of his unconventional small-wheeled bikes found that his swing-arm suspension was good at absorbing large bumps with frequencies up to 11 Hz, but not high-frequency 'road buzz.'[31] This is inherent in all suspension systems: The unsprung mass of the wheel is too large.

A 1 kg (2.2 lb) bicycle wheel cannot move up and down 40 times per second (cobblestones), much less 500 times per second (road buzz). This would mean accelerating the wheel vertically and then bringing it to a stop, before reversing the direction, up to 500 times a second. Shock absorbers cannot move fast enough for that kind of roughness.

Vibration is easiest to absorb near the source.

The energy of the entire wheel vibrating is too great to control, but high-frequency vibration can be absorbed at the tire level. Only a few grams of rubber and casing fabric at the tire's contact patch need to move – less than 1% of the weight of the front wheel. Such a small weight can oscillate quickly. And so the tire has little trouble following the contours of the bumpy road surface. That is why supple tires are far more effective at damping vibration – even those from cobblestones – than suspension forks. Suspension forks are useful for bigger bumps that the wheel encounters only a few times per second. Even cars use tires as the first line of defense. Springs and shock absorbers only absorb larger jolts.

There is another advantage to keeping the unsprung mass low: With little weight moving up and down, there isn't much energy to dissipate, so you don't slow down much as you ride over rough surfaces. Shock absorbers convert energy into heat, and that energy is no longer available to drive the bike forward. This is part of the suspension losses. These losses can be significant: The shock absorbers of off-road racing trucks get so hot that they are cooled with water. That heat is energy that is not available to drive the truck forward. Racing trucks have powerful engines and can afford to lose a few horsepower to heat their shocks. Yet even they use pneumatic tires as the first line of defense against vibration, leaving the water-cooled shock absorbers for dealing with big bumps.

The same logic explains why suspension stems or padded handlebar tape are even less effective at damping vibration than suspension forks. The entire front of the bike is moving up and down, making it virtually impossible to absorb the associated energy. That doesn't mean that padded bar tape is useless: It can reduce pressure points where the shape of your hands doesn't match that of the bars, and thereby improve comfort. Just don't expect it to damp vibration.

Big Bumps

Big bumps exceed the suspension travel of the bike's tires. Even a wide tire can only absorb relatively small bumps. For example, a 42 mm tire can handle bumps during normal riding that are about 30-35 mm tall. A 42 mm-tall bump will make the tire bottom out – something we try to avoid.

Expansion joints between road panels, frost heaves, roots that have warped the pavement, and the occasional pothole we cannot avoid, all create large, but infrequent, jolts. For all-road bikes, telescopic suspension forks add complexity and weight, plus they require frequent maintenance. Slender steel fork blades are a simple way to offer similar suspension.

The suspension of tire and fork form a system that works together. The tire handles the high-frequency, small-amplitude irregularities (road buzz, cobblestones), while the fork absorbs the bigger, but less frequent bumps. One without the other will produce sub-optimal results.

Suspension is useful for the bumps that are too big to be absorbed by the tires.

Front vs. Rear Suspension

Whether you experience more discomfort on your front or rear contact points depends on your power output and riding position. Riders with lower power outputs tend to sit more on the saddle, and thus benefit from rear suspension in various forms (suspension frames, suspension seatposts, cushioned saddles). Riders with greater power outputs don't sit on the saddle so much as hover over it, since they push down on the pedals with great force. For them, rear suspension is less important. Their more inclined riding position tends to put more weight on arms and hands, so they benefit more from shock absorption at the front of the bike.

This is why city bikes tend to have sprung seats and rear suspension, while performance bikes intended for rough terrain often have some form of front suspension. Rear suspension also tends to compress and extend in response to the pedaling forces. This bobbing motion can fight the rider's pedal stroke and reduce their power output, making the bike slower. Many mountain bikes with rear suspension have lockouts to prevent 'suspension bob' when climbing.

Casual riders put more weight on the saddle and benefit from rear suspension. Spirited riders have more weight on the handlebars and gain more from front suspension.

Contact Points

Contact points are where you touch the bike: saddle, handlebars, pedals. We've seen earlier that, by the time vibration reaches your body, the unsprung mass of your bike is too great to damp it. That is why padded handlebar tape and sprung saddles cannot reduce the discomfort from road buzz.

This doesn't mean that your contact points are not important. They determine how your body interfaces with your bike. If your saddle or handlebars don't match your anatomy, or if your pedals and shoes give you hot spots, you can suffer from discomfort or even nerve damage. Choosing your contact points according to your personal needs can make a significant difference in the comfort of your bike. Saddle, handlebars and pedals will be discussed in more detail in Part 3 of this book (see p. 208).

The fit of your contact points is important for the comfort of your bike.

Bike Fit

"The short reach to the handlebars makes my back ache slightly after 11 hours on the bike. A longer stem would be useful..."

A well-fitting frame is important for comfort on a bike, but the importance of a few millimeters of top tube length or stem height may be overrated. Part of the reason performance bicycles are so comfortable is that riders move around significantly as they change positions on the handlebars. The difference between the tops, ramps, hoods and drops is much greater than a few millimeters of stem length or height.

That said, it is important to be comfortable on the bike. The most stretched-out positions (on the hoods and in the drops) should not be so far forward that they are hard to reach. The saddle should have enough setback so that the rider is balanced over the pedals, and does not have to push against the handlebars to prevent sliding forward on the saddle.

How to figure out your very personal bike fit is a topic that goes far beyond this book. You will find some pointers in Part 1 (p. 14). This chapter explains how the frame's geometry affects how your bike will fit you. Most makers publish tables with frame geometries for the various sizes they offer. At first sight, these measurements can seem complicated – this chapter will help you understand how they affect the fit and ride of your bike.

Stack and Reach

The most accurate way to fit a bike is by looking at stack and reach. Stack is the vertical distance between the center of the bottom bracket and the top of the head tube. Reach is the horizontal distance between these two points.

Stack and reach tell you how tall and how long a frame is, providing the best way to compare two frames. The position of the saddle is adjustable, both up/down and fore/aft, so the rear part of the frame does not usually affect the fit. However, stack and reach are difficult to measure on a bike, since they do not correspond to the lengths of any tubes.

The most accurate way to size a frame uses stack and reach.

Top Tube

On bikes with horizontal top tubes, top tube length is easy to measure, and it tells you most of what you need to know about a bike's fit. Top tube length consists of two components: The reach is the portion that is ahead of the bottom bracket (see previous page), while the part that is behind the bottom bracket affects the setback. The seatpost and saddle also determine the setback, making setback relatively easy to change – either by sliding the saddle back and forth on the seatpost or, if more adjustment is needed, by swapping the seatpost for a model with more or less setback.

Setback is important because it balances the rider over the pedals. With the same top tube length and a steep seat angle, you have more reach and less setback, thus pushing the rider forward. This places more weight on the handlebars. With a shallow seat angle, the reach gets shorter and the setback longer. This requires the rider to bend more at the hips when they assume an inclined position for spirited riding.

Both frames have the same contact points and fit. The left bike has a longer seat tube and shorter (horizontal) top tube. The right bike has a sloping top tube and a sloping stem. To compare the actual measurements of the two bikes, you'll want to use reach and stack, not top and seat tube lengths.

Sloping vs. Horizontal Top Tubes

Most bikes used to have horizontal top tubes, because the long seatpost made necessary by a sloping top tube tends to flex from side to side, which can affect the bike's handling. The frame is well-braced, so it makes sense to make it large, and the overhanging seatpost short, to support the rider's weight optimally. Modern seatposts are stiff enough that this isn't a serious issue any longer. Some posts are designed to flex more back-and-forth than side-to-side. Choose a sloping or horizontal top tube based on your aesthetic preferences.

Stem

Stem length is added to the top tube length to determine the reach to the handlebars. Stem height also influences the rider's position. Together, these parameters determine how stretched out or how upright the rider is on the bike. Since the steerer tube is inclined, a higher stem position moves the handlebars rearward, and thus shortens the reach to the handlebars.

On a bike with a handlebar bag, overly long stems (>10 cm) are best avoided, because they push the handlebar bag too far forward. Keeping the bag close to the steerer axis reduces its inertia and keeps the handling nimble.

Raising the stem also shortens the reach.

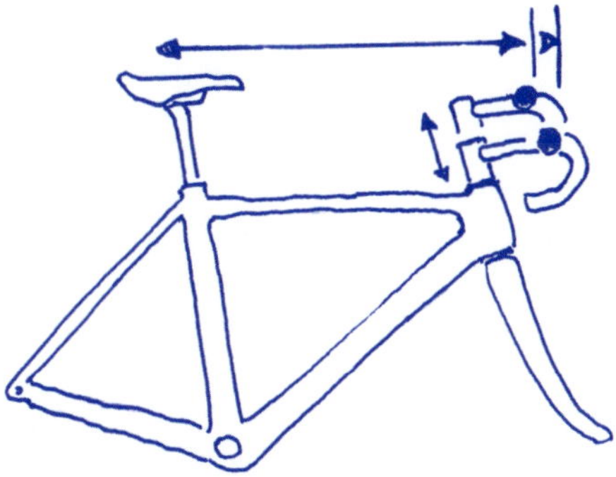

Seat Tube

Seat tube length is often used to designate the frame size, but this doesn't make much sense: Saddle height is adjustable over a relatively wide range. Modern frames often have sloping top tubes, or they extend the seat tube beyond the top tube, making this measurement even less meaningful.

Head Tube

With sloping top tubes, it has become easier to size frames by their head tube length than by their seat tube length. The important parameter is the stack: the handlebar height above the bottom bracket. Wheel size, tire size, clearance between tire and fork and fork design all affect the build height of the fork and thus head tube length. Still, when comparing two similar bikes, head tube length can give you a first approximation which bike is designed for a more upright or more inclined position.

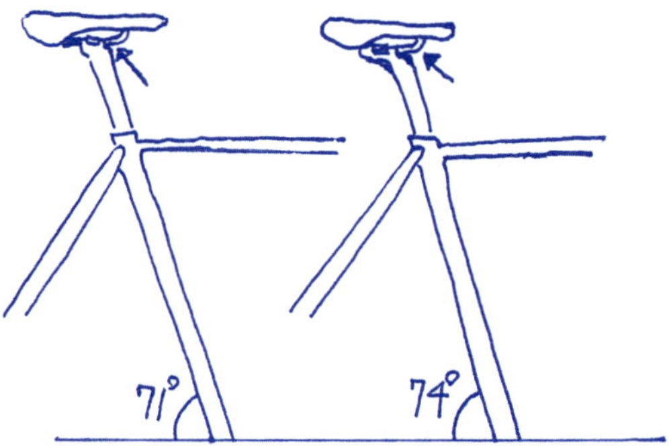

The same setback can be achieved with a shallow seat tube angle and straight seatpost (left) or a steep seat angle and an offset seatpost (right).

Seat Tube Angle

The seat tube angle is the angle between the seat tube and the horizontal. This angle helps determine frame setback. More important is saddle setback: how far the saddle is behind the bottom bracket spindle.

Seatpost design affects saddle setback. (Many modern seatposts have less setback.) The shape of the saddle rails also affects setback. (Traditional leather saddles often have less setback.) So a combination of steep seat tube angle, modern seatpost and traditional leather saddle offers the least setback.

The setback affects the cyclist's position over the bottom bracket spindle. Ideally, the rider is balanced so that little weight rests on their arms, yet their upper body forms an effective counterweight to the upward thrust of the pedaling forces. Small variations in setback do not appear to be important, because the rider's elbows are bent and their bottom can move around on the saddle, allowing the rider to find their optimal position on bikes with somewhat different geometries.

I have ridden and raced bikes with seat tube angles between 71.5° and 74° without noticing much difference. Other riders feel strongly that only certain seat angles work well for them.

Saddle Angle

A saddle that is tilted downward tends to make the rider slide forward, and they have to brace themselves with their arms. Obviously, the saddle should not hurt any soft tissues, but many riders can ride comfortably with their saddle's nose tilted slightly upwards. This anchors the rider in place, prevents sliding forward, and puts less pressure on the arms.

Standover Clearance

Some riders are concerned about standover clearance – being able to straddle the top tube with both feet flat on the ground. The Consumer Product Safety Commission (CPSC) even requires a minimum stand-over clearance of 1" (2.54 cm). On bikes with sloping top tubes, standover clearance is usually not an issue. But what about bikes with horizontal top tubes?

Quite a few riders feel that even a bike without standover clearance is no problem, as it is next to impossible to jump off a bike with both feet flat on the ground. When stopping, most riders put one foot down and lean the bike, so even a bike with no standover clearance does not jab us. I rode a bike without standover clearance for years without noticing, until a reader asked me about it after seeing a photo of me next to my bike. An exception is a tandem, where it is useful for the front rider to straddle the top tube with both feet firmly on the ground while the stoker gets on the back.

When stopping, riders only put one leg on the ground and lean the bike. This is safely possible even on a bike that does not have standover clearance.

Front-Center and Toe Overlap

Front-center is the distance between the center of the crank spindle and the center of the front wheel. Front-center, together with wheel diameter, determines whether a bike has toe overlap or not.

On bikes with toe overlap, the front wheel can hit your foot during tight turns. Not being able to turn the wheel unbalances the bike. Worse, as you pedal, the forward foot can end up on the inside of the front wheel. Then you can no longer straighten the front wheel, and a crash becomes difficult to avoid. On bikes with toe overlap, avoid pedaling while making tight turns at low speed.

On racing bikes, toe overlap can be acceptable. With narrow tires, the fork can turn quite far before the wheel touches your foot. And racing bikes are usually ridden at high speed, where the steering angle is small. On all-road bikes, wide tires can hit your foot even at small steering angles.

For the vast majority of riders, bicycles can be made without toe overlap. Toe overlap is harder to avoid on small frames, but a slacker head angle, more fork offset, and longer top tube (coupled with a shorter stem to keep the reach the same) can increase the front-center. Smaller wheels can be used to avoid toe overlap. There are no benefits and significant dangers to toe overlap, so it's best avoided through careful frame design.

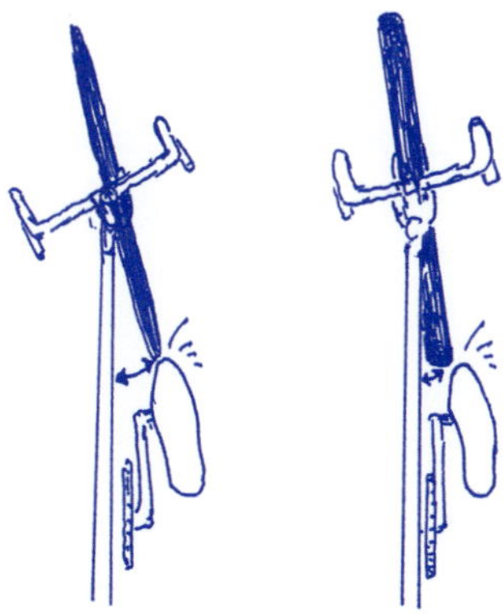

Toe overlap is a bigger problem with wide tires (right): A wider tire hits the foot at a smaller steering angle.

Riding Position and Power Output

To optimize comfort and speed, a cyclist's position should place only little weight on the rider's arms. Whether this results in a low or an upright position depends on the rider's power output, the strength of their core muscles, and the angle of their saddle.

The rider's upper body serves as a counterweight to the thrusts of the pedaling legs. Ideally, the two forces balance, and the rider's arms bear very little weight. The harder the rider pedals, the more their body should be inclined to provide a greater counterweight. A less powerful cyclist must sit more upright to obtain the same balance.

Every rider's power output varies over the course of a ride. That is why drop handlebars offer several positions. During lulls in the action, racers sit up and ride with their hands on the tops while spinning along. During strenuous efforts, they move their arms forward to the hoods or drops and lower their backs. They assume this more inclined position even when they ride in the draft of other riders, where there is no need to adopt an aerodynamic position. The racers incline their backs to balance the pedaling forces.

When a cyclist's position matches their power output, riding is comfortable for long periods. If the position is too low,

Drop handlebars allow riders to assume upright positions for casual riding and inclined positions for big efforts.

the rider must support their weight on their arms. This becomes tiring after only a few minutes. For this reason, coasting down a mountain pass fatigues the arms and shoulders more than riding on the flats: Without the pedaling action lifting the upper body, the arms have to support most of the weight.

An additional advantage of the inclined position on a performance bike is improved shock absorption. The rider's hips articulate to isolate the spine from road vibration and larger bumps. City bikes with upright positions often use sprung saddles or suspension seatposts for comfort.

Bent Elbows and Back Muscles

Experienced riders ride with their elbows slightly bent. This allows the arms to flex and absorb shocks. It also reduces the load on the arms and allows the strong back and core muscles to support the rider's weight. The stronger these muscles are, the more a rider can assume an inclined position comfortably. These muscles are trained by riding, which enables seasoned riders to assume a lower position more comfortably than novices.

Taking weight off the arms also improves the handling of your bike. If you lock your elbows to support your upper body's

Bent elbows isolate the rider from road vibration and decouple the bike from the rider's upper body movements.

weight on the handlebars, even small movements can throw the bike off course.

Bikes are self-stable, which means they continue on a given course without requiring any inputs. Good bikes require only a light touch to guide them where you want to go. Like good horses, bicycles react skittishly to heavy-handed inputs. Don't grip the handlebars hard, but gently guide the bike. It'll find its way almost on its own.

The one time to lock your elbows is when braking hard: Brace yourself against the handlebars to avoid flying forward.

Tall vs. Short Riders

A bicycle's head and seat tubes are inclined. This means that taller riders sit further back and put more of their weight on the rear wheel, while shorter riders sit more forward. To keep the front wheel loaded, larger bikes often use long stems while keeping the top tube proportionally shorter. The chainstays could also be lengthened to maintain the rider's weight distribution. However, the rearward weight shift of a tall rider is so pronounced that even a long stem and significant lengthening of the chainstays won't offset it.

Longer chainstays may blunt the bike's performance (see p. 65), so most makers just accept that a taller rider will have

Taller riders have more of their weight on the rear wheel.

more weight on the rear wheel than a shorter one. Since taller riders are usually heavier, moving their weight backward may actually be beneficial: It counteracts the greater wheel flop that would occur if the weight on the front of the bike was greatly increased.

Determining Your Position

How do you find out whether your position matches your power output and the strength of your back muscles? Try to bend your elbows when riding. You don't have to do it all the time, but you should be able to do it without feeling your muscles fatigue as they hold up your torso. If bending your elbows is hard work, then you have too much weight on your arms.

The different hand positions of drop handlebars let you gauge your overall position. If you find it comfortable to ride in the drops even during moderate efforts, you might consider a lower and further forward position for your handlebars. You will benefit from greater power in the lower position, and you still can use the hoods and tops of the bars when you are not going fast.

If you rarely use the drops, raising your handlebars might improve your bike's comfort and handling. You still will be able to get low by using the drops, but the higher positions will reduce fatigue.

Bike Fit Summary

Fit depends on many factors. Changing one variable means others also need to change to maintain a balanced frame. A bike fitter can help determine your fit. Bike designers create balanced frames for riders of different sizes and physiques. There is much more to frame design than can be touched upon in this overview.

Reliability

"After riding the 360 miles of the Oregon Outback over rough gravel roads, no bolts have vibrated loose, no parts have broken. The bike could turn around and do the event again. Its rider probably couldn't, though."

The secret to making a lightweight high-performance bike ultra-reliable is simple: use the right materials in the right quantities and in the right places.

For example, when the Bicycle Quarterly team started riding superlight steel frames, many predicted they would break, but even after half a dozen bikes have covered tens of thousands of miles each, we haven't encountered any problems. One reason is that a thinner tube flexes along its entire length, where a stiffer tube will concentrate the stresses at its ends.

Quality control becomes more important when using less material, because the tolerance for error is less. Lightweight frame tubes are much easier to overheat, so builder skill is of great importance.

Lighter parts tend to be more expensive, because the materials they use must be stronger. For example, many inexpensive aluminum cranks are cast, which is much weaker than forging. To compensate, the cheaper crank uses more material, so it's heavier than the forged crank. The extra material is often not enough to compensate for the inherent weakness of the manufacturing process, so the lighter forged crank will usually be stronger, too.

Racks made from aluminum rod are heavier and weaker than those made from chromoly steel tubing. With carbon fiber, voids in the material can reduce its strength. Lightweight parts made to high quality standards are much less likely to break than heavier, cheaper parts that may be full of voids. For example, it's rare for high-end carbon forks to break, but failures of inexpensive carbon forks are more common.

Good design is equally important. Poor design often leads to failures early in a bike's lifespan. Stress risers are places where sharp corners or transitions concentrate the stresses on a part. Some are difficult to avoid – where the seatpost enters the frame or where the pedal joins the crank – while others are the results of poor design.

Good design, well-chosen materials and high-quality constructions are key to making parts that are light and strong.

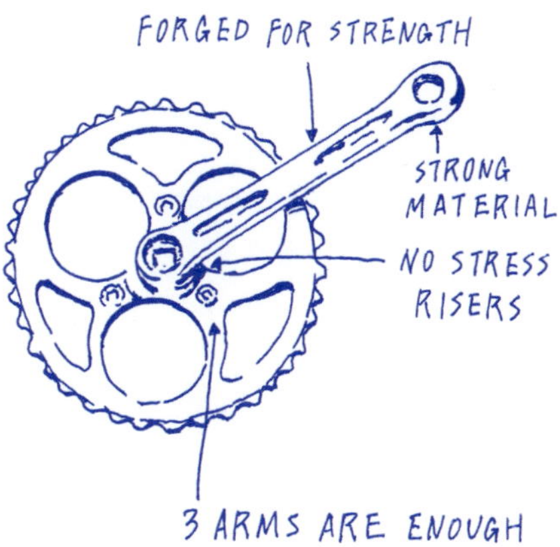

Experience and finite element analysis can predict these stress risers and help design parts that will last for a very long time, even under strong riders. Once identified, the stresses can be distributed more evenly. On carbon frames, seatmasts eliminate the long cantilevered seatpost that concentrates the stresses at its insertion point. The pedal interface is difficult to redesign, but cranks can be strengthened with additional material at the pedal eye.

If stress risers are not recognized, they almost invariably lead to premature failures. Most makers have enough experience to avoid failures on frames, but carbon forks and especially racks and fenders still break with alarming frequency.

These predictable failures can be avoided by observing the principles of sound engineering. For the customer, it can be hard to visualize the stresses without a detailed analysis. In addition to looking at the products, check the company's reputation. How many recalls has the company issued: Do they use good engineering and test their products carefully before introducing them, or do they use their customers as 'beta testers' and issue a recall after some of them get hurt?

Stress risers occur at abrupt transitions, for example, where the seatpost enters the frame.

A Quiet Bike

For many riders, a quiet bike is key to the enjoyment of cycling. Moving along effortlessly, with only the whir of the tires on the road and the wind in your ears, comes close to the bird-like feeling that we are trying to find on our bikes. Noises on the bike usually are a sign that something is wrong. Whether it's a squeaking chain, rubbing fenders or more mysterious creaks, they introduce an element of stress to the riding that is unnecessary. A good bike can and should be quiet.

Maintenance

Even the best bike will require periodic maintenance. Chains need to be replaced every few thousand miles to avoid premature wear on the other drivetrain components. Brake pads wear out, and so do tires. Rims and disc rotors also will need replacing eventually.

Beyond that, a bike should be mostly maintenance-free. On a well-designed bike, bolts should not come loose and parts should not break. A good bike should be able to cover the 1200 km (750 miles) of *Paris-Brest-Paris* without adjustment or repair.

Reliability Summary

Performance and reliability are closely related: Both are the result of careful design and high-quality construction. Modeling the stresses on a bicycle is difficult, and experience is usually the best way to predict failures and avoid them. Solutions and parts that have proven themselves are usually the best guarantee for a reliable bike.

Conclusion

Less Resistance and More Power

To make our bikes faster, we want to reduce the resistances that slow us down and optimize the frame so that our power output is maximized.

Of all resistances, rolling resistance offers the greatest gains, because the difference between slow and fast tires is so great. Reducing vibration through supple tires not only makes a bike more comfortable, but also faster: Less energy is wasted on suspension losses, and more of the rider's power is available to drive the bicycle forward.

Air resistance is large in absolute terms, but once you are on a performance bike and wear tight-fitting clothing, it is difficult to reduce your air resistance further. The importance of bicycle weight and bearing resistance is often overestimated.

Beyond these resistances, a bicycle's speed is determined first and foremost by how much power the rider puts out. Even on similar bikes, the same rider's power output can vary by 10% or more. The frame's flex characteristics can influence how much power the rider can generate. Riding a bike with optimized flex characteristics not only increases the rider's power, but also reduces the mental effort of maintaining a high speed: The rider's legs seem to spin on their own. Q factor has a similar effect: If the cranks are too wide, the rider's spin and power output will not reach their maximum potential.

Bike fit is determined by a variety of factors. For most riders, a few millimeters difference in reach or setback are of minor importance, yet a comfortable position is important for optimizing power output, especially over long distances. The rider's position should match their power output (and core strength). The more power a rider puts out, the lower (and thus more aerodynamic) their position will be. Drop handlebars enable the rider to assume different positions.

A bike's handling is determined mostly by its front-end geometry and the rotational inertia of its wheels. Different riding positions, loads and speeds require adjusting the geometry to obtain the same intuitive handling.

All these factors can work together to create a bicycle that is fast, comfortable and reliable. The beauty of modern all-road bikes is that it is no longer necessary to compromise between these goals: They are all related and optimizing one also improves the others.

Part 3

The Parts of the All-Road Bike

How do you turn the basic science of making a bicycle perform well into an actual bicycle? Part 3 examines how each part of the bike can be optimized for performance, comfort and durability. Even small parts matter: They all contribute to making an all-road bike a great machine that is far more than the sum of its parts.

"In a paceline, the all-road bike performed like a great racing bike. On gravel roads, it worked as well as a good mountain bike. On twisty descents, it outshone any other bike, since its wider tires offered more grip and better suspension."

Tires

"I was prepared for some improvement, but the magnitude surprised me. Obviously, the new tires rolled faster, but more than that, they changed the character of the bike."

Tires are the easiest and most effective performance upgrade for your bike. When getting a new bike, your first consideration should be the tires you want to use (see p. 13). Everything else will follow from that. In this chapter, we'll look at what determines the performance of your tires.

Pneumatic tires were the most important technological step forward in the entire history of the bicycle. Until the 1890s, cycling struggled to catch on because solid rubber tires gave a bone-shaking ride. Even hollow tires, still without pressurized air, did little to improve matters.

Pneumatic tires changed all that – not only were they more comfortable, but also much, much faster. However, pneumatic tires also introduced the bane of cyclists' existence: flat tires. In the 1890s, changing a tire required removing the side of the rim with dozens of screws. Most cyclists dealt with a puncture by walking to the next train station and taking the train home, where they could disassemble the wheel, fix the tire, and reassemble it.

When pneumatic tires became available, the cycling world split into two groups: long-distance tourists and scorchers. The former prized their self-reliance, so they stuck with airless tires and rode slowly and deliberately. The scorchers, on the other hand, rode fast on their pneumatic tires, but never strayed far from home for fear of getting stranded.

Tire technology soon improved. Once punctures could be fixed on the road, cyclotourists also adopted pneumatic tires. They saw their reach vastly increased and their enjoyment multiplied. Combining the speed of the scorchers with the long-distance adventures of cyclotouring gave birth to a new sport – randonneuring – and new bikes that were the ancestors of today's all-road bikes.

Over the next century, the joyful revolution of riding on a cushion of air was forgotten, and cycling (almost) returned to the harsh old days. Racers adopted narrower and narrower tires, pumped to higher and higher pressures, until their bikes (almost) emulated the boneshakers of old. Many touring

Pneumatic tires revolutionized the comfort and speed of bicycles.

cyclists, trying to avoid flat tires at all costs, adopted puncture-resistant tires with belts and foam layers under the tread that stopped even thumb tacks from penetrating the air-filled inner tube. These tires were so stiff that they resembled the 'hollow rubber' of the old days, and they robbed tire and bike of much of their liveliness.

Recently, these trends have been reversed. Racers are moving to wider and wider tires, and running them at lower pressures – not only on the cobblestones of Paris-Roubaix, but even on the smooth roads of the Tour de France. Riders have found that by adopting wider tires, the risk of flats is much reduced, and even casual riders can enjoy supple high-performance tires without worry. Once again, cyclists enjoy the speed of scorchers and the independence of tourists.

How important are tires? Imagine going at moderate speeds of about 15 mph (25 km/h) on a popular tire model. Change your tire for the fastest model we have tested, and you immediately go 18 mph (29 km/h). That is 17% faster, with the same power output. It's hard to believe, yet it's been confirmed in numerous tests.[32] At higher speeds, you will still see a speed difference of 8% or more. (Compare that to aero wheels, which give you a speed advantage of about 1-2% at high speeds.)

In a peloton, wind resistance is greatly reduced, and your tires are the main resistance you have to overcome. The advantage of faster tires is greatly magnified, allowing riders on faster tires to get more rest between pulls at the front.

Sheltered in a group, faster tires make an even bigger difference than when riding alone.

What Makes a Tire Fast?

For Bicycle Quarterly's tire tests, we evaluated the performance of more than 35 different tire models, both in roll-down tests and with a power meter. Rather than a simple ranking of tire brands and models, our goal was to determine what makes a tire fast.[33]

Our research confirmed what professional racers have known for decades: To make a fast tire, you need a supple casing. Everything else is secondary: width, tread compound, etc. The five fastest tires in our testing ranged in width between 24.5 and 49 mm. They included tubulars and clinchers. The common trait was that they all used extra-supple casings.

Supple Casing

A supple casing is the defining feature that differentiates a fast-rolling, comfortable high-performance tire from a slow and harsh-riding utility tire. What makes a casing supple?

Casings are made of threads. The finer the threads, the more supple the casing. Coarse threads are much stiffer, and take more energy to flex. The threads of a utility tire are more than twice as thick as those of a high-performance tire. With oversize frames, increasing the tube diameter by 12% changes the feel of the bike – imagine what happens if you double the diameter of the threads in your tire casing!

The actual material of the threads also plays a role. Traditionally, silk tubulars were prized over cotton ones, because silk is more supple. Today, only a few handmade tubular tires still use silk casings. Cotton tires still exist, but polyester is the most common thread material, and it can offer excellent performance as well. Within each material, there are different grades, and it is difficult to generalize which material is superior.

A finer thread means that more threads are used, and so 'threads per inch' (TPI) has become a shorthand for casing suppleness. This is only partially correct. Finer threads do result in more threads per inch of casing fabric, so the TPI increases. However, if you cram as many threads as possible into each inch – thus getting the highest TPI – you create an ultra-dense weave that makes the casing stronger, but also less supple. The fastest tires use the finest threads, but in a slightly looser weave that makes them even more supple (and coincidentally decreases their TPI). The looser weave also decreases the tire's strength, so there is a limit beyond which the cut resistance of the casing is no longer adequate.

Some companies advertise improbably high TPI of 150 or more. Threads that fine are too fragile to make bicycle tires – these companies count each layer of the three casing layers, so their 300 TPI tires are the equivalent of what most manufacturers count as a 100 TPI casing.

Finer threads are more fragile and have to be handled with greater care during the manufacturing process. Factories that usually make budget tires are poorly equipped to handle such delicate casings, so they coat them in thick rubber to make the casing more resistant to damage. The extra rubber negates much of the effect of the thinner threads. As so often, focusing on a single number – TPI – is misleading, because it cannot capture the subtleties involved in making a high-quality product. The most supple tires have fine threads, a moderately dense weave, and only as much rubber coating the sidewalls as is needed to protect the threads from abrasion.

Tire Width

The reason why racers have moved to 25 mm-wide tires is simple: They are faster. Bicycle Quarterly's testing found that 25 mm-wide tires roll faster than 23 mm ones, which in turn are faster than 20 mm tires.[34] One reason is that the contact patch of a wider tire is shorter. At the same pressure, a wider tire deflects less as it rotates. This reduces its hysteretic losses. And if you run the wider tire at lower pressure, the suspension losses are smaller. Either way, resistance is reduced.

On smooth pavement, there is no speed benefit to tires wider than 25 mm, but also no disadvantage. Our testing found that even 53 mm-wide tires do not roll slower than narrower tires, at least at speeds of 29.5 km/h (18.3 mph).[35] In these tests, we compared tires with the same supple casings, just different widths. Conversely, a wide touring tire with a stiff casing will roll much slower than a supple racing tire of any width.

While tire width above 25 mm doesn't make much of a difference on smooth pavement, on rough roads wider tires roll faster, because you can run them at lower pressures. This reduces the suspension losses that make up the bulk of tire-related resistance on rough pavement (see p. 42).

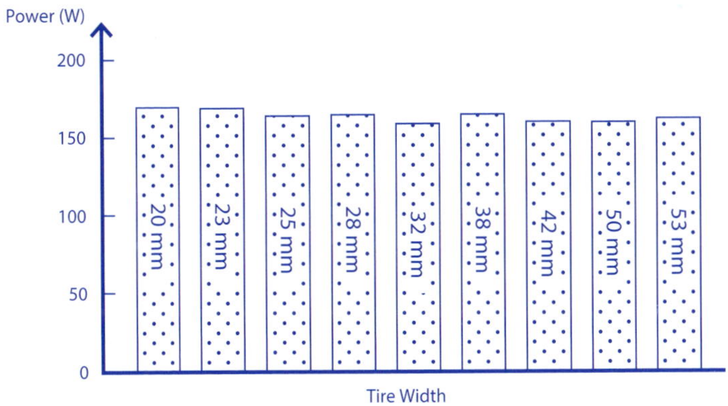

Power at 29.5 km/h

Power to pedal at 29.5 km/h with tires ranging in width from 21 to 53 mm on smooth pavement. 21 and 23 mm are slowest; there is no statistically significant speed difference among the other widths. (All tires same casing.)

Tread Thickness

A tire's tread is the rubber that touches the road. A thinner tread makes the tire more flexible and thus faster, because it requires less energy to flex as the wheel rotates. The effect is small, but significant. In our testing, a set of well-used 27 mm-wide tires with 1-1.5 mm of tread worn off was 2.1% faster than a new set of the same tires.[36]

With wide tires, tread thickness has less effect on speed, probably because the larger radius makes the tread less convex and easier to flex. When we tested brand-new 42 mm-wide Rene Herse Extralight tires and compared them with a set that had been ridden more than 6500 km (4000 miles), we found no significant difference in speed.[37]

For narrow tires, tread thickness is a compromise between speed and longevity. Since tread can be worn down safely only to a thickness of about 1 mm, a 2 mm tread will allow for only 1 mm wear. A 3 mm tread allows for 2 mm of wear, and thus will last twice as long. Making the tread thicker doesn't offer much benefit. Since the tire wears faster in the center than on the shoulders, wearing more than 2 mm off the tread leaves a square profile to the detriment of handling.

A narrow tire's tread is more convex than that of a wide tire. The narrow tread takes more force to flex, unless it is very thin.

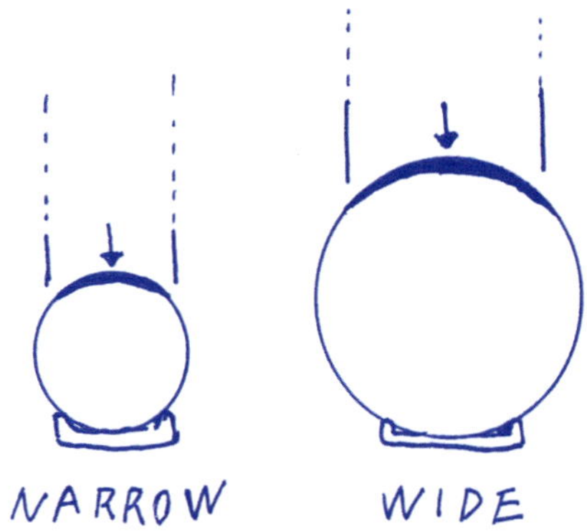

Tread Pattern

Tread blocks can squirm and cause significant energy losses. Both height and width of the tread blocks determine how much they squirm. Small blocks squirm more, whereas interconnected treads or negative grooves don't slow down the tire much.[38]

Some dual-purpose tires have knobs that are large enough so they don't squirm, yet small enough so they dig into mud and soft soil for traction. By arranging the tread knobs so there is always the same amount of rubber in contact with the road surface, even when the bike leans into a corner, these knobby tires can offer predictable grip on pavement, too. In effect, these tires function like slick tires with a negative tread – the negative portion of the tread is so large that only the knobs remain of the original slick surface. Designed carefully, these tires retain most of the qualities of slicks on pavement *and* function like knobby tires on loose surfaces.

For racing tires, a fine tread may increase performance, since it allows the tire to conform better to the road surface without having to flex the entire tire casing.

Weight

The fastest tires are light because they use thin, supple casings and relatively thin tread rubber. However, weight by itself does not determine performance. Some very light tires perform poorly. Heavy tires rarely perform well, since extra material doesn't just add weight, but also makes tires stiffer.

Tires with folding beads are lighter than those with steel beads, but there appears to be no discernible difference in performance. For a discussion of how the weight of tires and wheels affects a bike's performance, see page 52.

Tire Pressure

Tire pressure was discussed earlier (see p. 44). Consistently, we have found that higher tire pressure does not improve performance. Choose a pressure that feels good, and your bike will be as fast as it can be.

Tires with supple casings require slightly higher pressures than stiff tires. On stiff tires, the sidewalls provide much of the support that prevents the tire from collapsing during fast cornering. In extreme cases, the tire sidewalls can support the entire bicycle without any air pressure. Supple tires rely almost exclusively on air to hold up the bike. Imagine the tire casing as a rubber spring and the air inside as an air spring. The two springs work together. If you use a supple casing that is less stiff, you need to compensate by increasing the pressure.

Despite their higher inflation pressure, supple tires are more comfortable and roll faster, because their casings absorb less energy as they flex, and because they transmit less road shock, which reduces suspension losses.

On rough surfaces – gravel, cobblestones, mud – low tire pressures offer better performance. The softer tires absorb road irregularities better, and the larger contact patch improves traction. On loose surfaces like gravel, hard cornering is impossible, and sidewall collapse is less of a concern.

Riding on smooth asphalt requires the same power output, no matter how hard you inflate your tires. (Data for supple 32 mm tires.)

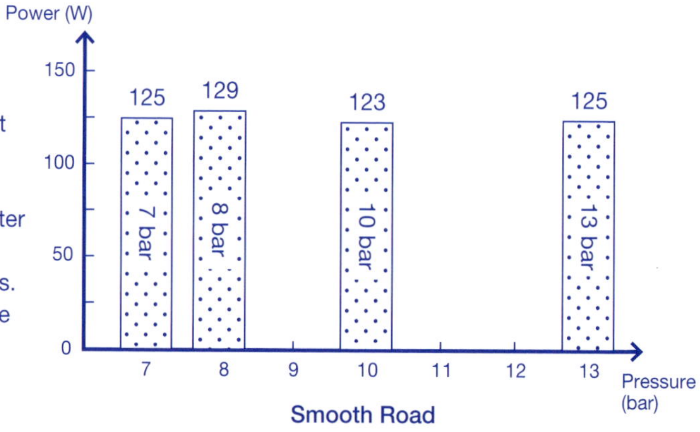

Finding the right pressure

Experimentation is the best way to determine the optimum tire pressure. The pressure written on a tire's sidewall is only the maximum safe pressure – it's not the pressure you should use.

Your tire pressure needs to be high enough to prevent the tire sidewalls from collapsing during hard cornering. If your pressure gets too low, you'll first notice that your bike is running wide in corners. It will also tend to follow ruts in the road more than usual. If you lower the pressure further, your sidewalls can suddenly collapse during hard cornering.

You also want to avoid bottoming out on big bumps, as this can cause pinch flats (with tubes) or cut the tires (tubeless). The minimum safe pressure is determined by your weight, riding style and surface conditions. Some makers list minimum pressures for their tires, but they have little meaning beyond indicating that the optimum pressure lies within a broad range. Light cyclists can usually ride at less than the recommended minimum pressure.

Above the lower limit dictated by safety, you can choose your tire pressure to dial in the feel of your bike. Do you want a little give in your tires to offer beneficial flex (see p. 68)? Or do you prefer a stiffer feel?

Wider tires have a greater range between the minimum and maximum safe pressures. With quality tires, pressure does not change your bike's speed, only its feel. Experiment with different tire pressures to tune the feel of your bike.

Tubulars vs. Clinchers

On most surfaces, tubular tires are slightly slower than clinchers.[39] However, the effect is much smaller than previously thought. Tubular tires have higher hysteretic losses, possibly because the tire creeps on the rim. However, the round profile of the tubular allows the casing to flex around the tire's entire perimeter, whereas a clincher can flex only on three sides, as the fourth side is made up by the rigid U-shaped rim. This reduces the tubular's suspension losses, and it compensates for much of the higher hysteretic losses.

For narrow and medium-width tires, the shock absorption of tubulars equals that of a clincher that is about 10% wider – assuming similar casing construction. To obtain the comfort of a 25 mm tubular, you need a 28 mm clincher. However, with very wide tires – above approximately 40 mm – the rim makes up much less of the circumference of the tire, so tubular tires lose their advantage.

Tubulars remain popular in road racing due to the lighter weight of their rims and their greater traction. In cyclocross, they used to offer a great advantage because they are much less likely to pinch-flatting, since their rims lack the sharp horns that hold clincher tires on their rims. Modern tubeless technology has eliminated the risk of pinch flats, making clincher tires a viable choice even in cyclocross. Pro racers, who are limited by UCI rules to a maximum tire width of 33 mm,

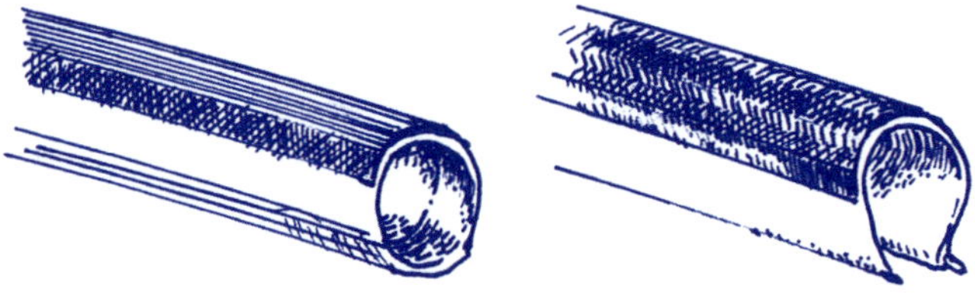

A tubular tire (left) is sewn around the tube. It is glued onto the top of the rim.
A clincher tire (right) is open against the rim. It interlocks with the rim sidewalls.

still choose tubulars for their better shock absorption and traction. For amateur cyclocross racers, 38 mm wide supple clinchers offer similar performance and shock absorption.

With tubular tires, high tire pressure is *slower* than lower pressure on all road surfaces. A tubulars' performance relies to a greater degree on low suspension losses. Higher pressure increases suspension losses and reduces the tubular's performance. To optimize performance, tubulars should run relatively low pressure.

Tubes

Thin, lightweight tubes probably reduce the rolling resistance of tires, but the effect is small. In our testing, the difference between ultra-thin and standard tubes was not statistically significant. However, high-quality latex tubes roll significantly faster than butyl tubes.[40]

Tubeless

Tubeless tires eliminate the tube altogether. The tire sits tightly on a specially designed rim that forms an airtight seal with the tire. True tubeless tires must have air-tight sidewalls, which requires an extra rubber membrane that makes the tires stiffer and slower than standard tires.[41]

The thin sidewalls of supple tires are not airtight. However, sealant can be used to make supple tires airtight, so they can be run tubeless. Sealant also has the advantage that it seals small punctures, thus preventing many flats. If a puncture is too large to be sealed, the rider will have to install an inner tube.

Eliminating the tube makes the tire more supple, but liquid sealant inside the tire adds friction. The two effects cancel each other. Setting up tires tubeless is useful for rough terrain where it eliminates the risk of pinch flats, but it's not faster than running butyl tubes, and slower than using latex tubes.[42]

Temperature

Warm tires roll faster, because rubber becomes softer at higher temperatures. This effect is very significant, with speed differences of 8% or more between warm and cold days.[43] More importantly, grip is also much-reduced at temperatures below 10°C (50°F). If you are trying to set a record, your chances of success are higher on a warm day.

New vs. Broken-In

New tires can have higher resistance than tires that have been broken in for a few hundred kilometers. The difference amounted to about 3.5% on one particular tire with a relatively coarse tread pattern.[44] It is likely that the edges of the tread blocks are worn smooth, thus decreasing their resistance. It is also possible that the threads in the casing loosen up a bit and thus reduce the resistance of the tires.

Acceleration

Wider tires inevitably are heavier than narrow ones. The difference is smaller than many cyclists imagine, because the casing of high-performance tires is thin and does not weigh much. The tread rubber of wider tires can also be thinner, since the wear is spread over a larger surface. The beads are the same for tires of any width. The weight of the extra air inside is negligible. Wide tires with supple, lightweight casings offer much greater air volume for a small weight penalty.

This means that wider tires accelerate only very slightly slower than narrow ones (see p. 52 for a discussion of wheel weight and acceleration). Furthermore, the extra weight of wider tires can be compensated by using smaller wheels, so the rotational inertia remains the same: That way, a wide tire can accelerate as fast as a narrow one.

Tire Pressure and Power Output

With wide tires, tire pressure can be used to tune the bike's flex characteristics and maximize the rider's power output (see p. 58 on 'planing'). Wide tires will perform and feel different depending on their pressure. Experiment with tire pressure to figure out what works best for you.

Cornering

Wide tires have much more cornering grip. Two factors are at play here. First, the wider tire simply puts more rubber on the road. This means that there are more interlocking interfaces between tire and road surface, and thus more grip. That is why race cars use wide tires.

Wider tires also run at lower pressures, so they adhere better to the road surface, rather than skipping over irregularities like a narrow high-pressure tire. When a narrow tire loses contact as it skips, it momentarily breaks traction. This reduces the cornering limit of the tire. If the tire tread is always in contact with the road surface, grip will be maximized. This effect is especially pronounced on rough and bumpy roads.

Casing suppleness also plays a role in cornering traction. A supple tire will absorb more irregularities, and thus keep the tread in better contact with the road surface.

Wider tires have more cornering grip.

Rim Width and Cornering

The width of the rim changes the profile of the tire. With a wider rim, the tire becomes U-shaped rather than O-shaped, and, in theory, the sidewalls can support greater loads. This is said to improve cornering.

However, the sidewalls of supple tires are not stiff enough to support significant weight. The extreme example are tubular tires, which are completely O-shaped. Professional racers run tubular tires in part because of their superior handling – further evidence that with supple tires, rim width does not affect handling.

Supple tires rely almost exclusively on air pressure to resist cornering forces. Supple tires must be inflated a bit harder to make up for the reduced casing stiffness (see p. 142), but even stiff tires can collapse during hard cornering, if they are not inflated enough: As the tire compresses, it always becomes more O-shaped, reducing its 'spring rate' and making it easier to compress it further.

The optimal pressure for cornering is low enough to keep the tire from skipping over bumps, but high enough that the sidewalls don't collapse.

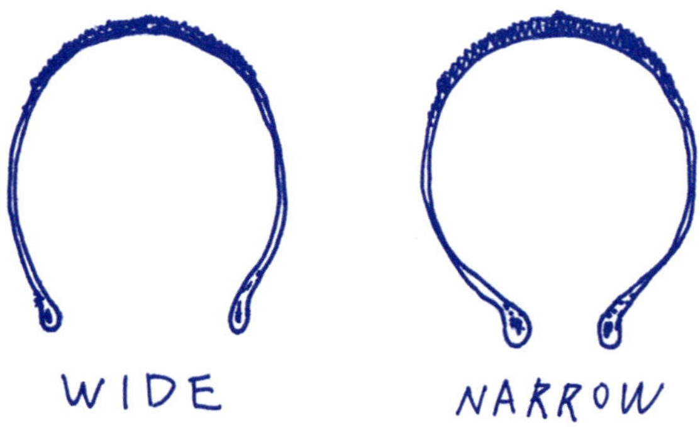

On a wide rim (left), a tire tends to be U-shaped, which keeps the sidewalls relatively vertical. The same tire on a narrow rim (right) is O-shaped and less stiff vertically.

Optimizing Tire Tread

Tire tread has numerous functions, yet it is often misunderstood. On car tires, tread patterns mostly serve to evacuate water from the contact patch to prevent hydroplaning. Bicycle tires don't hydroplane, and in the past, many believed that the most logical tire tread pattern was none at all – a perfectly slick tire. Even today, many tire manufacturers consider tread patterns mostly ornamental.

This overlooks that a well-designed tread pattern can interlock with the pavement and increase cornering traction.[47] This effect is especially pronounced in the wet, but it is important even on dry roads. The best tread pattern has as many small edges as possible to interlock with road surface imperfections. The traditional chevron or file tread pattern found on many classic high-performance tires works very well in this respect.

Why are fine ribs used on bicycle tires, but not for cars or motorcycles? Car tires are square, because they don't lean into corners. Their entire surface is in contact with the road at all times, and ribs would wear away quickly. Motorbikes have too much power to make fine ribs practical. They weigh much

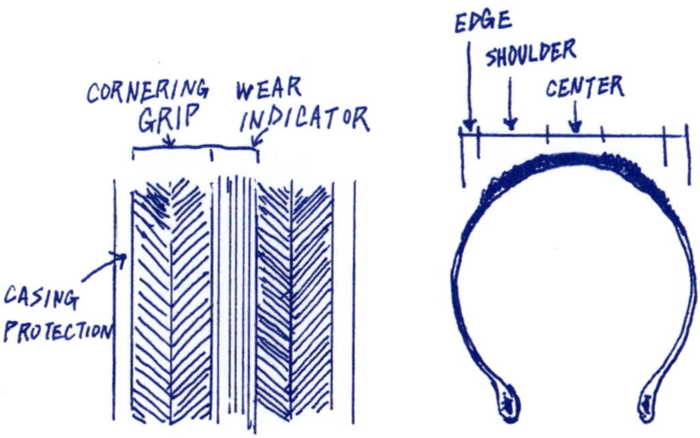

Different areas of the tire tread serve different purposes.

more than bicycles, so their tires are pushed into the road surface and interlock more firmly with the pavement. On bicycle tires, the shoulders that touch the road during cornering do not experience significant wear. Fine ribs improve cornering grip and last as long as the tire.

An optimized tread pattern is divided into several parts. In the center of the tire, the tread wears, so it doesn't serve much function beyond acting as a wear indicator. When the bike leans over during cornering, the tire rolls onto its shoulder. Traction is crucial in this situation, and the tread pattern here should be optimized for interlocking with the road. Beyond the shoulders, the edge of the tire does not touch the ground, but the casing still is exposed to sharp rocks and other debris. Here, a smooth tread protects the casing from damage.

Gravel, Mud and Snow

On loose gravel, tire tread serves little function. The tire does not slip on the stones that it touches, but the stones themselves slide on those underneath. Tire tread cannot increase the friction between rocks that the tire does not touch, so it makes no difference on loose gravel.

On soft dirt roads, the entire tread digs into the surface (leaving tire tracks), and provides enough friction to provide traction. Tread patterns play an even greater role on mud and snow. Here, smooth treads slide on the slippery surface, and knobs that dig deep into the substrate offer significant traction advantages. Treads for mud and snow have the knobs spaced widely, so that the tread does not clog up, which would render it ineffective. Maintaining a high speed is best on these surfaces, because it allows the tire to self-clean as the dirt is pulled away from the tires by the centrifugal force.

Tread patterns that are optimized for mud or snow often work poorly on hard surfaces and pavement, since the knobs squirm to the detriment of rolling resistance and cornering traction. Rolling resistance and grip can be improved by

Why race cars have slick tires

If tire tread is designed to interlock with the road surface, why do race cars and race motorbikes have slick tires, unless racing rules require them to have tires with tread? Doesn't this imply that tires without tread have the most grip?

Racing tires for cars and motorcycles use a very soft rubber, so the road surface gets imprinted into the tire, forming an interlocking bond between road and tire. The downside of the soft compound is that these tires wear out quickly. Race cars often need several tire changes during a single race.

Few cyclists would want to replace their tires every few hundred miles, so bicycle tires use harder, more wear-resistant rubber that cannot conform as well to the road surface. That is why slick bicycle tires offer less grip than is ideal.

spacing the knobs more closely, but this is not ideal. Tires with closely spaced knobs may have an 'adventure' look, but they tend to offer the worst of both worlds: In mud and snow, they clog up and perform as poorly as a road tire. On pavement, they squirm like a dirt tire and run wide when cornering.

Careful design can overcome these problems, by making the knobs large enough that they don't squirm and spacing them wide enough that they don't clog up with mud. The tread pattern can place the knobs so the tire rolls seamlessly from one knob to the next, both when moving forward and when leaning into a corner. Knobby tires with this design can offer predictable cornering and low rolling resistance on smooth pavement *and* excellent traction in soft mud.

Some knobby tires arrange the tread knobs so that the frequencies of the noise they generate overlap and cancel each other. These tires run much quieter on pavement than conventional knobbies.

On ice, no rubber tread can get any purchase, and steel studs are the only option to improve traction. Overall, knobby tire treads offer their greatest advantages in soft mud and snow. When you can see the tire tread's imprint in the surface, you know that the knobs dig deep enough to improve traction.

Knobby tires improve traction on surfaces where the tread pattern leaves an imprint in the surface.

Tread Rubber

Many tire companies used to believe that harder rubber compounds had lower hysteretic losses and thus offered better performance. On real roads, the opposite appears to be true. Softer rubber compounds tend to roll faster, because they conform better to the road surface, thus reducing the suspension losses. Skate boarders have long known that softer wheels roll faster.

In the past, grip and durability were mutually exclusive. Cyclists had a choice between soft rubber that offered good grip, but wore out quickly, and hard rubber that lasted a long time, but tended to slip when cornering, especially in the wet. In recent years, rubber technology has improved greatly. The best tires today combine excellent grip with good longevity. Generally, black tread rubber offers the best grip. Colored treads tend to offer less adhesion, especially in the wet.

Flat Resistance

Tires with thin, supple casings are more flat-prone, because they are easier to penetrate by sharp objects. On paved roads, debris rests on the hard road surface, and when you ride over it, the only place it can go is into the soft rubber of your tire. If the debris is sharp, it will get stuck, and with each wheel revolution, it will get hammered further into the tire, until it punctures the tube.

When pneumatic tires first became popular in the 1890s, horseshoe nails were the bane of cyclists. Today's cyclists fear the tiny wires from the steel belts in car tires. When a semi-trailer truck gets a flat, the driver often doesn't notice until the tire has been shredded, spewing wires over the road for miles. Other common causes of punctures are glass as well as freshly crushed, sharp-edged rocks that are spread on wintry roads to provide traction on snow and ice. Goathead thorns are common in some regions, and they seem to puncture most tires.

Passing cars continuously stir up road debris, tossing it around until it comes to rest on the edge of the road, where few cars disturb it. That is why flats are almost guaranteed when riding on the shoulders of busy highways. Quiet backroads, where cyclists ride in the main lane that is swept clean by cars, are relatively clear of debris, and the likelihood of flatting is much reduced. (Plus, riding on small roads is much more pleasant.)

To avoid flats, you can attack the equation from two sides: reduce the likelihood that debris gets stuck in the tire, or prevent it from getting through the tire. Punctures usually don't happen immediately. Debris merely gets stuck in the tread during the first pass, but then it gets hammered into the tire with each revolution until it punctures the tube.

If the debris is removed after the first pass, most flats can be avoided. Racers used to wipe their tires with their hands after riding through glass or other road debris. Rather than use your hands and risk injury (especially on the rear, where your hand can get stuck between tire and frame), you can use tire wipers (also called tire savers). These are small wires positioned to rub lightly on the tire tread. This brushes off most debris before it has a chance to lodge itself in the tire.

Debris on highways is moved time and again until it comes to rest on the shoulder. Flats are almost guaranteed when riding there.

Once the debris is stuck in the tire, it usually is only a matter of time until it punctures the tube. Tire manufacturers use steel and Kevlar belts in an attempt to prevent these punctures. The belts work fine for glass and thorns, which tend to get pulverized if they cannot penetrate the tire quickly. However, steel wires and sharp rocks eventually work their way through the puncture-resistant belt. It may take a little longer until the tube is punctured, but the result is the same. The only solution is to increase the thickness of the tire tread until most objects are too short to make it through. To prevent puncture-resistant tires from becoming unbearably heavy, makers often use a layer of foam under the tread, so that obstacles get stuck before they reach the inner tube.

Puncture-proof belts and foam layers make tires less supple. Ride quality and performance suffer. Bicycle Quarterly's testing found that puncture-resistant tires roll about 15 percent slower than supple high-performance tires.[48] Over the course of a day-long ride, you will be over an hour ahead on faster tires. Even if you have to fix a flat every day – and few people experience that many flats – you will arrive sooner on faster tires. More important, your bike will feel much more alive and comfortable on supple, fast tires.

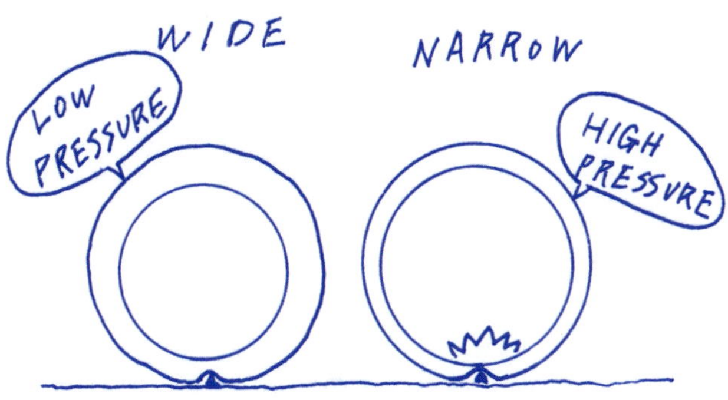

Low pressure, wide tires roll over debris that punctures hard, narrow tires.

One of the best ways to reduce flats is to run wider tires. Wider tires are inflated to lower pressures, so they roll over debris that would puncture a narrower, harder tire. A wide tire pushes less hard onto the debris, so it's less likely to puncture. This makes it possible to ride wide tires with supple casings and without puncture-resistant belts in environments where narrower high-performance tires would flat much too often.

A third solution is to accept some punctures as inevitable and use sealant inside the tire to seal small holes. The sealant used in tubeless tires will seal holes formed by small pieces of glass, steel wires and even goatheads.

When riding on gravel or dirt, punctures are rare. The rubber of your tire is harder than the ground, and debris gets pushed into the ground instead of into your tire. That is how cyclocross racers can ride on relatively fragile hand-made tubular tires. Off-road, sidewall cuts from sharp rocks are your biggest concern. Riding technique plays a significant part in avoiding these sidewall cuts. Beyond the obvious precaution of trying to avoid sharp rocks, sitting lightly on the bike allows the tires to deflect off the rocks rather than get cut by them. It's also important to run relatively low tire pressures. A soft tire is far more difficult to cut than a hard one.

Another issue is the 'pinch flat.' When hitting an obstacle, a tire can compress so much that the inner tube gets crushed between rim and road surface. In most cases, pinch flats occur as two punctures side by side ('snakebites'), caused by the two

When hitting an obstacle, the tire can compress so much that the tube gets pinched between the rim and the ground.

horns of the rim crushing the tube as the tire bottoms out. If the rim is bottoming out on an irregular-shaped object or while the bike is leaning into a corner, you might get only one hole.

On paved roads, pinch flats are becoming a thing of the past as riders move to wider tires, which can take bigger hits without bottoming out. A pinch flat indicates that your air pressure is too low. An immediate remedy is pumping your tires to a higher pressure. In the long term, it makes sense to use wider tires if your bike can fit them.

On gravel, especially when racing in a peloton, it can be difficult to avoid large rocks that can pinch-flat even wide tires. Tubeless tires are resistant to pinch flats (there is no tube to pinch). However, tire and rim can still get damaged if the wheel bottoms out too harshly.

Avoid Re-Flatting

"I had five flats on a single ride and never found what caused the punctures." When riders report multiple flats on a single ride, they are often caused by the same object. If the debris that caused the initial puncture is not removed, then it's only a matter of time until the new tube punctures as well. Debris stuck in the tire tread can be very difficult to find.

During one 600-kilometer randonneur brevet, I flatted every 80 kilometers (50 miles). After the third flat, I finally found a tiny piece of wire embedded in the tread. The wire was so short that it protruded only when the tire tread flexed under load, so I did not notice it when I checked the tire with my hands. Yet on the road, it nibbled away at the new tube with each wheel revolution, until the next puncture occurred.

Always look for the debris that caused the puncture – or at least the cut it left in the tire, because some debris punctures the tube without getting stuck. If you cannot find the cause of the flat, replace both the tube and the tire, if you carry a spare. At the end of the day's ride, with good light and ample time, it will be easier to find the object that caused the puncture.

Why Wider Tires?

The realization that wider tires roll no slower than narrow ones has revolutionized bicycle design. In the past, it seemed impossible to build a bike that could keep up on fast road rides *and* handle rough roads. Narrow high-performance tires were known to offer superior comfort and performance, but they were too fragile to be used every day, much less for touring.

With wider tires, these limitations and compromises went away. By putting wider tires on a performance bike, you can get the speed of a racing bike with the rough-road versatility of wide tires. Add a low-rider rack on the front, and you have a bike that you can take touring, too. Yet even loaded, you still enjoy the spirited feel of a performance bike. And since wider tires run at lower pressures and thus have fewer flats, you can use supple high-performance tires for all your rides. Wide tires also last significantly longer, because the wear is spread over a larger tread surface, which makes expensive high-end tires very affordable on a per-mile basis.

This is truly a win-win-win situation, and it's changed the way we ride. A typical ride might now include paved climbs, gravel roads across the top of the mountains, and exhilarating descents on pavement again. The new breed of all-road bike climbs like a racing bike, handles gravel roads as well as a mountain bike, and offers more cornering grip during the descents than either of them.

How Wide is Too Wide?

If wider tires are better, how wide can you go? The biggest constraint is the need to fit the tire between the chainstays of the bike. Road cranks are narrower than mountain bike cranks, which helps facilitate a good spin (see p. 70), but they leave less space for wide tires. The practical limit for road bikes appears to be around 55 mm. Dropped chainstays and One-By cranks can open up another 5 mm or so.

Even that might be a bit wider than ideal for many riders. Our research has shown that 55 mm-wide tires roll as fast as narrower ones, but they are more sensitive to tire pressure. Their large, soft sidewalls tend to be more affected by cambers and longitudinal depressions in the road surface. Very wide tires also make more noise.

For riding on paved roads, 38-42 mm-wide tires are entirely sufficient. They make packaging the tire between the bike's chainstays much easier, and they are less sensitive to pressure. If you wonder why professional racers ride narrower tires, you will find the answer on p. 23.

At some point, tires become too wide to fit between the chainstays and cranks of a performance bicycle.

The Downsides of Wide Tires

We've explored the many advantages of wide tires, so what are the disadvantages? Remarkably few. Wider tires are inevitably heavier, but not by much, and not enough to make a difference in performance – provided you run supple high-performance tires. By choosing smaller wheels, the rotational inertia remains unchanged, and thus a bike with wider tires will accelerate as quickly as one with narrower tires, and handle similarly in corners, but with more grip.

The biggest drawback may be aesthetics. If you love the look of a skinny-tire racing bike, then wide tires will not appeal to you. There is nothing wrong with that – we all choose our bikes in part based on appearance. Supple casings make a bigger difference in the feel and performance of a bike than tire width. A racing bike with narrow, but supple, high-performance tires will be as fast and almost as comfortable on paved roads as an all-road bike with wider tires.

Narrow tires are part of the aesthetic appeal of racing bikes.

Tires Summary

Tires narrower than 25 mm are quickly becoming obsolete: They are slower, less comfortable and offer less grip in corners than wider tires. Above 25 mm, tire width by itself has little influence on smooth-road performance. On rough roads, wide tires will be faster because they reduce suspension losses. Since few cyclists ride exclusively on smooth roads, most of us will be faster and more comfortable on wide tires. Wide tires are also more flat-proof and last longer, making it feasible and cost-effective to use supple high-performance tires for everyday riding.

Wheels

"The front wheel, with its high-profile rim, transmitted a little more shock than the box-section rim we also tried on the bike."

The performance benefits of high-end wheels are often overestimated. Wheel weight seems to matter comparatively little for the bike's performance (see p. 52), and aero wheels decrease a bike's wind resistance by only 2-3% – less than lowering your stem by 20 mm. Lightweight wheels do change the feel of the bike, especially when you rock it from side to side while sprinting or climbing out of the saddle. Light wheels can be strong, too. Well-built wheels should withstand 30,000+ kilometers (20,000+ miles) of riding on rough roads without needing service.

Spokes don't break from overloading, but from fatigue as the spoke is loaded and unloaded when the wheel rotates.

The wheel flattens at the bottom, which unloads the spoke at 6 o'clock. With each wheel revolution, every spoke passes through that spot, where it is slightly detensioned, and then tensioned again. Over time, that causes the spoke to fatigue.[45]

To get the maximum life out of your spokes, you want the detensioning to be as small as possible (as a percentage of the overall spoke tension). That is what double-butted spokes are for: Thinner in the middle, they stretch more, which means that they don't detension as much as a thicker spoke would. Yet the ends, where spokes break if they fail, are thick and strong. It may seem counterintuitive at first, but the thinner mid-sections make double-butted spokes more durable than thicker straight-gauge spokes. And of course, they are lighter and more aerodynamic, too.

Wheels tend to go out of true when you hit a bump and a spoke detensions so much that it goes slack. As the spoke is tensioned again, the nipple unwinds a bit. Now the spoke has less tension, so it will go slack more often, allowing the nipple to unwind more and more until the spoke is completely slack.

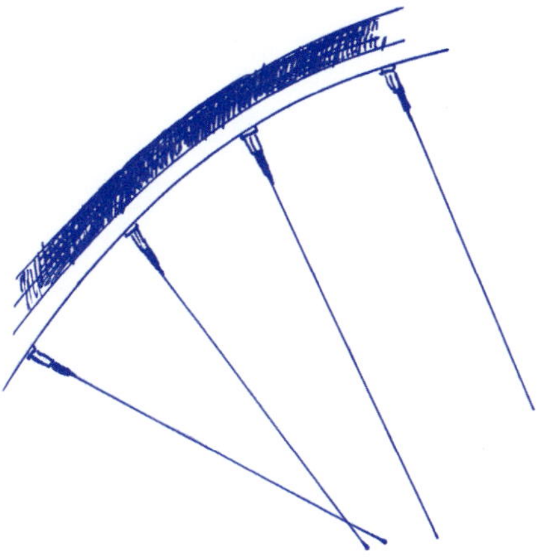

Double-butted spokes are not just lighter, but also stronger.

How Many Spokes?

How many spokes do you need? For many years, 36 spokes were the standard, then most bikes went to 32 as modern spokes became stronger. Wider tires transmit fewer shocks to the rim, and with smaller 650B rims, the spoke bracing angle is greater, which makes the wheel stronger. That means that 28 spokes are plenty, even for rough roads, and even 24-spoke wheels have proved reliable.

With disc brakes, the left hub flange is offset to the inside to make room for the rotor. The left spokes have a steeper bracing angle, making the 'dished' wheel less strong. Stressing the wheel further, the entire braking loads are transmitted by the spokes (unlike on a rim brake wheel).

The rear wheel never sees significant side loads, which means that the smaller spoke bracing angle (to make room for the freewheel/cassette) doesn't matter as much as it would on a front wheel. Even so, a few extra spokes won't hurt. Wind tunnel studies indicate that the rear wheel is in such turbulent air that its aerodynamics may not matter much, and the little extra weight isn't a big deal, either.

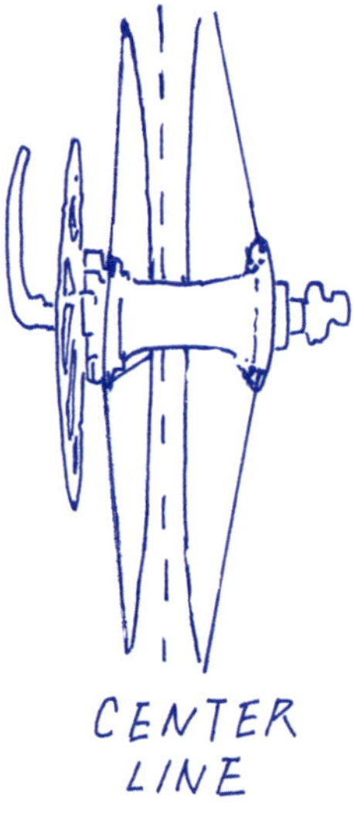

To make room for the rotor, disc-brake front hubs have offset flanges and steeper spoke bracing angles on the left side (seen from above).

Rims

Most rims today are stiff and strong. Deep-dish rims are stiffer, but also provide a harsher ride.[46] They also catch crosswinds, making the bike challenging to ride on windy days.

Carbon rims can be lightweight and strong, but they don't work well with rim brakes, because they don't dissipate the heat generated by braking. For bikes with disc brakes, they are fine, and only their high cost is a disadvantage. For rim brakes, aluminum rims remain the best choice.

If rims crack, it's usually caused by poor design or substandard materials. Carbon rims can suffer from voids in the material. Once you've eliminated those problems, what you want from a rim is a good fit of the tire. Classic rims should seat the tire automatically as you inflate it. The tire shouldn't come off even if you have a sudden blowout. With tubeless tires, the fit needs to be even more precise, so the tire seals easily and doesn't blow off the rim despite lacking a tube to reinforce the joint between tire and rim.

The rear wheel is weaker and more stressed than the front wheel.
A few extra spokes go a long way toward durability.

The Frame

"Steel, titanium and carbon all can be used to make great bicycles."

The frame is the heart of the bicycle. It is far more than just a structure to connect the wheels to the pedals, handlebars and saddle: The frame determines the interaction between rider and bike. It defines the feel and personality of the bike. We've covered frame flex and 'planing' as well as frame geometry and handling in earlier parts of this book. This chapter is about the materials used to make the frame.

Frame Materials

For much of the 20th century, most performance bicycles were made from steel. During the 1990s, titanium and aluminum became popular. Today, most high-end bicycles are made from carbon fiber. Bamboo and even wood have emerged as credible alternatives to the traditional materials. Great bikes can be built from all these materials. Each has unique advantages and disadvantages, and each has found its place in the bicycle market.

Steel

Steel benefits from more than a century of experience in making high-performance frames. Through decades of racing on road and track, through millions of miles ridden on tours and adventures, steel frames have been fine-tuned to the nth degree.

Steel is relatively inexpensive and also relatively easy to work with. That means it can be used to make budget frames, yet it also can be worked by skilled craftspeople to create some of the best-performing custom bikes.

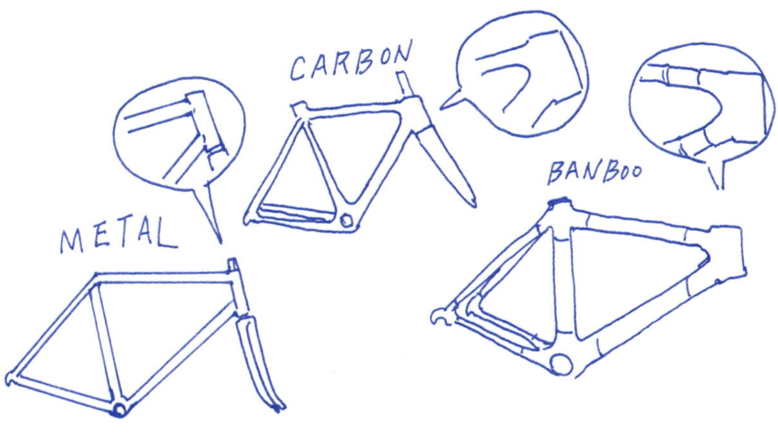

Frames with very similar ride characteristics can be made from metal – steel, titanium, aluminum – carbon and even bamboo.

Steel is available in many shapes and forms. Steel frame tubing exists in a far greater selection than it does in titanium or aluminum. Steel is easily shaped, which is useful for making all-road bikes. Curving chainstays around wide tires, or raking forks with extra offset to adjust for the pneumatic trail of wide tires is relatively easy – just bend it a bit more.

For a fully equipped randonneur or cyclotouring bike with integrated fenders, custom racks and other features, steel remains the best material. The slightly higher weight of the frame compared to titanium or aluminum is not as important as it may seem at first sight: The frame makes up only a small portion of the complete bike's weight, and once the rider's weight is factored in, it becomes insignificant.

The grade and quality of the steel tubes does not affect a frame's ride quality – all steels have the same modulus of elasticity. The grade is important for the frame's durability: The thinner the tubing walls, the higher the grade of the metal needs to be. Thinwall steel tubes are heat treated to increase their strength, ductility and elongation properties (but not their stiffness). Just as important is the manufacturing quality of the tubes: Imperfections can cause the tubes to break prematurely.

Beyond the quality of the tubes, it's the framebuilder's skill that determines a frame's durability. Experienced builders know how to braze or weld the joints completely without overheating the steel. Overheating weakens the metal and often causes premature failures.

Titanium

Titanium is as strong as steel, but weighs only half as much. (It is also half as stiff – see below.) Titanium does not corrode, so it does not require paint to protect it from the elements or even salt water. These properties make it appealing for bicycle frames.

For a long time, titanium was an almost mythical material. Titanium alloys strong enough for engineering applications were first produced in the Soviet Union. The material was so scarce outside the Eastern Bloc that the United States maintained a strategic stockpile for use in spyplanes and supersonic aircraft. The first titanium bikes were made by Speedwell in Britain in 1973, but titanium only became widely available and (somewhat) more affordable with the collapse of the Soviet Union in the 1990s.

Titanium requires great care during welding to avoid contamination with oxygen. That, and the high cost of the material itself, makes titanium frames expensive. There are no real economies of scale: Making ten titanium frames costs almost ten times as much as making a single one. That is why titanium is used mostly for high-end custom frames.

Aluminum

Aluminum is the lightest metal used for making bicycles. In theory, aluminum frames can be lighter than titanium or steel ones, but the limited fatigue life of aluminum means that frames without a large margin of safety tend to lack durability. This inbuilt margin of safety also means that many aluminum frames are relatively stiff, leading to a reputation for a harsh ride.

The first high-performance aluminum bikes were introduced in 1936 by Nicola Barra, a small custom builder in Paris. His welded aluminum frames were prized by randonneurs and racers, including several riders in the 1949 Tour de France.

Aluminum frames saw a resurgence during the 1970s, and by the 1990s, exotic aluminum alloys like scandium rivaled titanium and carbon fiber in light weight and performance. Today, titanium has taken over the high-end market for metal frames, and aluminum is mostly reserved for lower-end bikes where mid-range carbon fiber is still too expensive. The factories in Asia that make most production bikes have automated the welding of aluminum frames, making aluminum frames the cheapest frames today. This has limited the material's appeal at the higher end of the market.

Comparing Steel, Aluminum and Titanium

All three metals found in bicycle frames have the same ratio of stiffness to density. Titanium is half as dense as steel and half as stiff. For aluminum, it's one-third. To keep things simple, let's compare steel and titanium – the same applies to aluminum.

If you made identical frames out of steel and titanium, the titanium frame would weigh half as much, and it would be half as stiff. If you want to make the titanium frame as stiff as the steel one, you can double the wall thickness of the tubes. Now you have two frames that look the same, weigh the same and have the same stiffness. The only advantage of the titanium frame is that the thicker tubes are more dent-resistant.

There is a better way to increase the stiffness of the titanium frame: Increase the diameter of the tubes. This uses only a little more material, but it increases the stiffness to the level of the steel frame. Now you have a titanium frame that is as stiff as the steel one, but weighs less.

Why can't we increase the diameter of the steel frame, too, and make the walls 50% thinner than those of the titanium frame? The steel walls would get so thin that they'd easily dent or even buckle. They wouldn't provide enough support for braze-ons. There is a minimum wall thickness for frame tubes. For steel, it is somewhere around 0.4 mm. That is why steel frames tend to be heavier than titanium frames.

Does this mean that an aluminum frame will be even lighter? In theory, yes. In practice, aluminum has smaller margins of safety: It breaks, rather than bends. To prevent aluminum frames from flexing and breaking, they tend to be overbuilt. This has given them their reputation for being stiff. If stiffer frames were faster, we'd all be riding aluminum frames!

With steel, even the lightest tubes can be stiff enough for tall and strong racers like Andy Hampsten, who won the Giro d'Italia on a frame made from tubes with ultra-thin 0.4 mm walls. If that frame was stiff enough for Andy Hampsten, then most steel frames may be too stiff for smaller and/or less-strong riders. With steel, you cannot reduce the wall thickness further, and making the diameter smaller might compromise the durability of the joints. With titanium, it is possible to make a frame that is more flexible than even the most flexible steel frame – which may be what some smaller riders need.

Carbon Fiber

Carbon fiber is different from metals. The direction of carbon fibers resembles the grain of wood: Carbon fiber's strength varies depending on the direction of the load. In theory, this allows fine-tuning the characteristics of the frame exactly as you want.

Carbon fiber is the newest material for making bicycles. High-strength carbon fiber was developed during the 1960s in Japan. In 1976, Alan in Italy created the world's first production carbon bikes. The company was already making frames that used aluminum tubes bonded into aluminum lugs. Alan replaced the aluminum tubes with Japanese carbon ones.

Today, most frames for high-performance bikes are made from carbon fiber. At the top end, carbon allows making lighter frames than other materials. Mid-range carbon fiber bikes offer light weight and excellent performance at affordable prices due to their efficient production.

Carbon fiber works best in large uninterrupted shapes – that is why it's a great material for airplane wings. Early 'lugged' carbon bikes sometimes suffered from failures at the joints. They were replaced by more swoopy shapes created in molds. That still leaves problems where the wheels, seatpost, headset and other parts attach. On carbon forks, the crown area that wraps around the tire is highly stressed. The smaller the opening for the tire, the easier it is to make a smooth transition around it. With wider tires, more material must be used, so carbon forks for wide tires used to be significantly heavier than those for narrow tires. As the design of carbon bikes has matured, standards have changed to suit the new material. For example, tapered steerer tubes create a smooth transition to the large fork crown needed to wrap around a large tire.

The directionality of carbon fiber is both its strength and its weakness. When a carbon bike falls over, the impact can hit the fibers in the direction of their greatest weakness. The carbon can crack or even shatter, whereas most metals would flex and either spring back to their original shape, or, in the worst case, suffer a small dent.

Carbon fiber requires costly molds, but once the tooling exists, the cost per frame is relatively low. The skills for laying up the carbon layers in the mold are easier to learn than welding or brazing metal frames. This makes carbon ideal for production bikes, where the mold costs can be amortized over large production runs and per-unit costs are low.

Carbon fiber comes in different grades with different stiffnesses and strengths. Standard modulus is the least expensive, used almost exclusively on entry-level frames. Intermediate modulus is the strongest carbon, so it accounts for most of the carbon in high-end frames. High and ultra-high modulus is stiffer, but also more brittle. It is used sparingly to stiffen high-end frames without adding weight.

Carbon fiber relies on a good bond between the fibers for ultimate strength. The resin that bonds the carbon fibers does not contribute to the material's strength, so makers want to use as little as possible to reduce weight. During the early days of carbon fiber frames, this could lead to the carbon being too 'dry,' with insufficient bonding between layers. Even today, voids can reduce the strength of carbon frames, forks and components.

Bamboo and Wood

Bamboo has similar properties to carbon fiber tubes, and can be used to make similar bike frames. The tubes are wrapped with carbon fiber at the ends, creating strong joints similar in concept to the lugged joints of classic steel bikes. Just like carbon fiber, bamboo can be used to make bikes that offer excellent performance.

Sano-Magic in Japan makes high-performance frames out of wood. The aged mahogany imbues the frames with many of the qualities of carbon fiber – including ultralight weight – with the added advantage of being able to shave the frame tubes to fine-tune the flex of the bikes. Downsides are the very high cost and the difficulty of accommodating wide tires, since wood tends to be more bulky for the same strength and weight.

Manufacturing

Different frame materials require different manufacturing techniques, which makes them uniquely suitable for different bicycles. Metal frames are assembled on adjustable jigs, making it relatively easy to build custom frames. Carbon fiber is made in molds, and even custom frames are assembled from pre-made parts, limiting the customization. For example, until recently, curved carbon chainstays were available only for cyclocross bikes with 700C wheels. On 650B bikes, the curves of the chainstays didn't line up with the spot where the tire required the most space.

Steel is relatively inexpensive, so more time and effort can be spent on custom-built racks and perfectly integrated lights and fenders, yet the cost of the best custom steel bikes remains comparable to top-of-the-line bikes made from other materials. With titanium bikes, the expensive material and complex manufacturing process account for much of the high cost.

Comparison of Frame Materials

Steel:
+ easy to shape
+ mature technology
+ inexpensive
- frame tends to be a little heavier

Titanium:
+ can be made lighter and/or more flexible
+ does not rust
+ does not dent easily
- expensive

Aluminum:
+ less expensive
+ can be very light
- must be stiff for good durability

Carbon fiber:
+ almost anything is possible
+ can be very light
- requires expensive tooling
- easily damaged

Ride Quality

You may have heard about the 'lively ride' of steel, the 'harsh ride' of aluminum and the 'flexy feel' of titanium. Most of that is caused by frame design rather than the material itself. It is possible to make bikes with very similar feel from all materials. Four of my favorite bikes are made from thinwall steel (Rene Herse), titanium (Firefly), aluminum (Alan cyclocross) and carbon (Open U.P.P.E.R., not mine). On the other hand, I have ridden steel frames made with thick-wall tubing that felt more like oversized aluminum and offered none of the lively feel that make my favorite bikes so much fun to ride.

In theory, there are small differences in vibration damping, especially between metals and carbon fiber, but in practice, tires and other components make a bigger difference in how a bike feels. The characteristics of a great frame can be obtained from all materials. Choose the one that you like best.

Unconventional Frames

The basic design of bicycle frames has not changed in more than 130 years. Frames still consist of two triangles, hence the name 'diamond frame.' It made sense to make frames that way when they were made from metal tubes. Carbon fiber allows almost unlimited freedom in frame design, and some makers have tried to come up with alternatives to the classic diamond frame.

None of the alternatives were successful, and today's superlight carbon bikes still have a top tube, down tube, seat tube and two sets of stays, just like most bikes since around 1890. One reason is that the Union du Cyclisme International (UCI) that governs bicycle racing has mandated diamond frames for professional racing, but that does not explain why gravel and adventure bikes also have turned away from innovative frames. It appears that the diamond frame is well-understood, and makers know how to balance the frame stiffness to make frames that perform well.

The Fork

"I rode the same course on consecutive days, with the same loads, first with one fork, then with a second fork that has different blades. There was a noticeable difference in comfort. The second fork did not take the edge off the bumps like the first one did."

The fork is an important part of a bike, but its role is often misunderstood. The fork is the only part of a frame that isn't triangulated – the blades are supported only at one end. This allows the fork to flex as the bike hits bumps. The fork is also where most of the braking forces occur and are transmitted to the frame. And, of course, the fork steers the bike.

Steering Precision

You may have heard that a stiffer fork makes a bike handle with more precision. This assumes that the fork sees significant side loads, liked the suspension of a car. This isn't the case: The bike leans as it corners, and the forces are aligned with its axis. There are no significant side loads on the fork during cornering, otherwise, the bike would fall over. There are small side loads when the cyclist initiates a turn, but even forks with flexible blades are stiff enough to resist these loads.

Fork vs. Wheel Flex

Side loads occur when you ride out of the saddle. As you rock the bike from side to side, side forces load up the fork and the front wheel. Even then, the front wheel flexes more than the fork. Modern wheels with few spokes are especially weak laterally, yet they corner fine, because neither wheel nor fork stiffness have a noticeable effect on how a bike corners.

Even during hard cornering, the bike is balanced and there are no side loads on the fork.

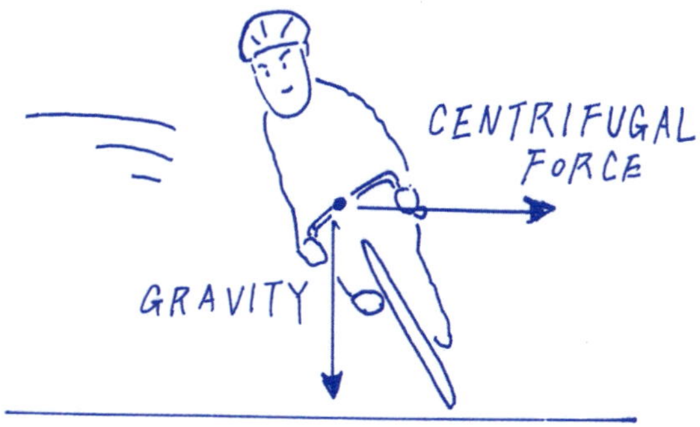

Comfort

Some fork flex is beneficial for comfort: The fork is part of the bike's suspension. A well-designed fork flexes enough to damp shocks that are too large for the tires to absorb. The diameter of the blades, head angle, fork offset, and location and radius of the rake determine how much a fork flexes.[49]

Most steel forks are raked at the bottom. This bend makes the fork more flexible in the vertical plane. The best steel fork blades taper to a very small, round diameter, further enhancing the ability to flex and absorb shocks.

Caliper mounts for disc brakes stiffen the fork at the bottom, so this portion can no longer act as suspension. Even with disc brakes, the upper part of the fork blades can be designed to provide some shock absorption. On carbon forks, too much flex would delaminate the carbon layers, so these forks tend to be stiffer than the best steel forks.

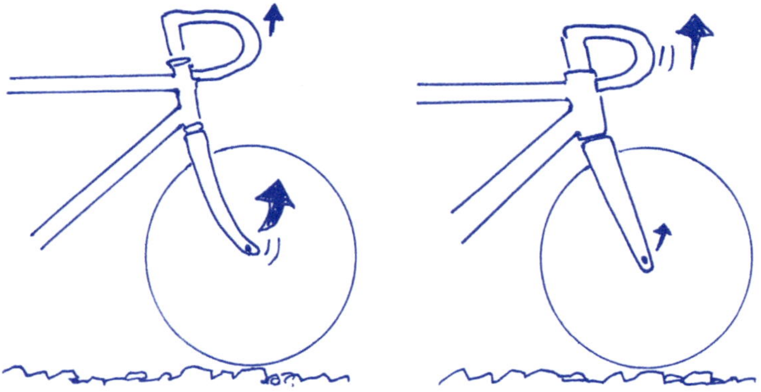

A curved fork (left) has a shallower angle at the fork ends, allowing the fork to flex upward to absorb shocks. A straight fork (right) deflects less and transmits more shocks to the handlebars.

Speed on Rough Terrain

On rough terrain, a stiff fork is not only less comfortable, but also slower. More shocks transmitted to the handlebars mean greater suspension losses that slow the bike. In Bicycle Quarterly's testing, both a suspension fork and a steel fork with flexible blades required 7% less power to pedal over rough ground than a very stiff fork.[50]

Mountain bikers have known this for a long time: Suspension makes a bike not only more comfortable and easier to control on rough terrain, but also faster. For all-road and gravel bikes, most bumps are small enough that complex and heavy suspension forks do not appear to be necessary, but a small amount of undamped suspension built into the fork does improve comfort and speed.

Shock-absorbing forks can reduce the power required to pedal the bike on rough terrain by 7%. (Data for 19.3 mph/31 km/h)

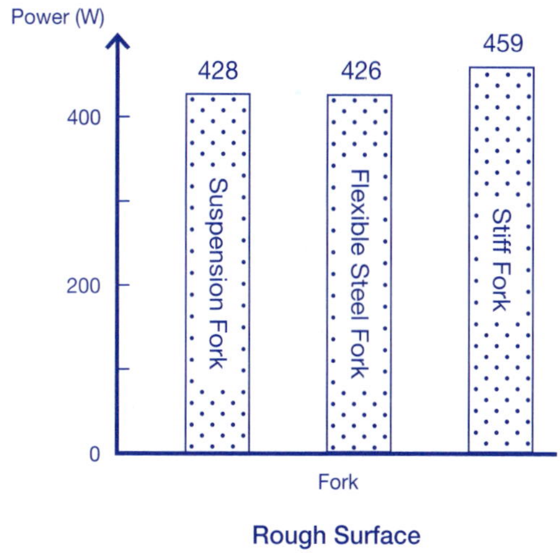

Brake Judder

Bikes with cantilever brakes can judder when braking. The brake cable hanger is usually attached near the stem, while the brakes are on the fork blades. Brake forces compress the fork. Braking changes the bike's weight distribution and puts a greater load on the front wheel, so the fork bows forward.

If the fork blades are too stiff and the fork crown/steerer tube too flexible, the flex occurs in the upper part of the fork. This effectively shortens the distance from the cable hanger to the brake and releases the brake. Now the front of the bike moves upward again, which straightens the steerer tube and lengthens the distance from the cable hanger to the brake, tightening the brake again. A rhythmic oscillation starts: The brake tightens and releases in quick succession, causing the fork to judder.

The best way to avoid fork judder is with properly designed forks. If the fork blades are relatively flexible and the fork crown and steerer tube are stiff, the fork will not judder. Quality brake components with very little play in the pivots can also reduce or eliminate judder on some bikes.

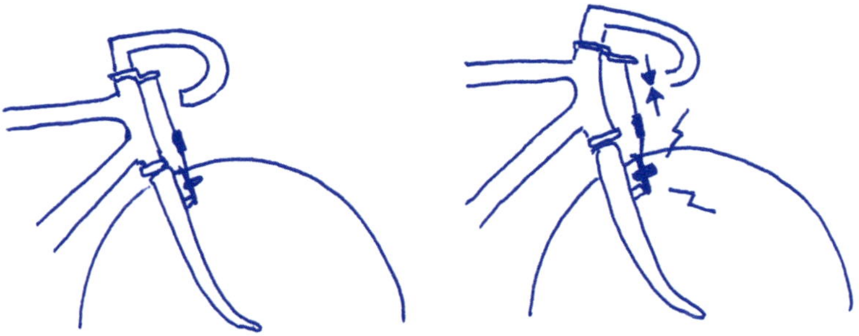

Brake judder is caused by rhythmic flex in the steerer tube that tightens and slackens the brake cable.

Racks

Front racks that connect to the fork crown and the dropouts can stiffen the fork considerably. When carrying a heavy load, this is less noticeable, as the added weight increases the deflection of the tire. When riding a bike with such a front rack unloaded, the reduced comfort will be noticeable. Low-rider racks can be removed when riding 'empty.' Porteur racks can be attached to mid-fork braze-ons, allowing the lower portion of the fork to flex and absorb shocks.

Fork Summary

The most important quality of a fork is its ability to absorb shocks. Most forks are stiff enough to resist the (small) sideloads that occur when the rider initiates a turn. With rim brakes, the brake forces occur at or near the fork crown. Disc brake forks need to be stiffer, because the entire braking force is transmitted through the fork blades.

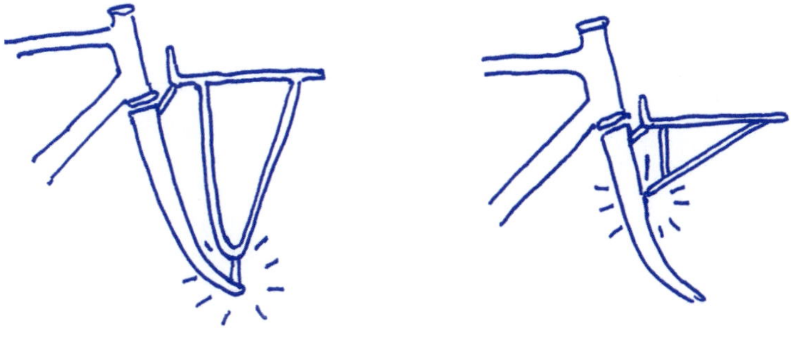

Racks stiffen the fork (left), unless they leave the lower fork blades free to flex (right).

Brakes

"The descent is perhaps the best one yet. We get into a wonderful rhythm and really enjoy becoming one with our bikes and the road. My front brake bites hard as I pull on the lever, decelerating and increasing traction on the front wheel as I turn in. Then I slowly release the lever as I round the corner, to let the bike fly onto the next, short straight. This game repeats itself dozens of times, and my smile gets bigger and bigger."

When you fly down a mountain descent and brake hard for a turn, you need two things from your brakes: stopping power and modulation. The stopping power slows you down, while the modulation allows you to meter out that power with precision. Both are equally important; they are somewhat related.

Braking power is important only on the front brake. During hard braking, the rear wheel barely touches the ground (if at all) and cannot transmit much braking power. The rear brake is useful when traction is limited on slippery surfaces, or to dissipate heat during prolonged braking.

Rims Brakes vs. Disc Brakes

Rim brakes and disc brakes are fundamentally the same, because rim brakes in effect are disc brakes. On a rim brake, the rotor also carries the tire, but the general idea is the same: Two pads squeeze a disc to slow down the wheel. The same basic rules apply to both systems.

The main difference is that rim brakes use very large rotors (the rim), which adds power, but the brake calipers have to reach around the tire, which increases flex. Disc brakes have smaller rotors (less power), but the calipers can be stiffer (less flex). Advantages and disadvantages of each type are discussed below.

Some bikes are equipped with drum or roller brakes. The enclosed mechanisms of these brakes suffer from overheating and don't work well on performance bikes, so they are not discussed here.

Braking deep into turns requires excellent brakes.

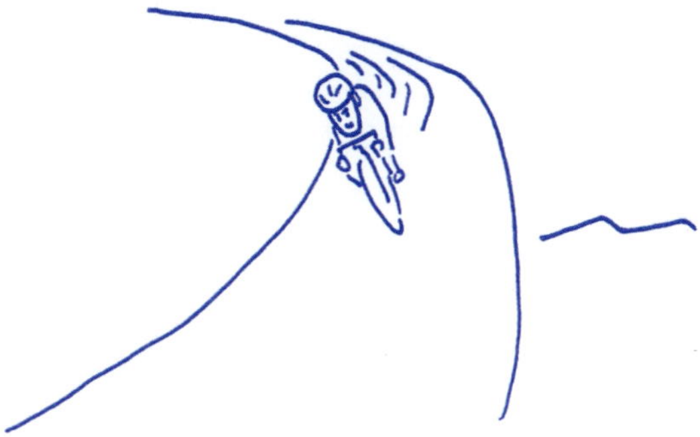

Mechanical Advantage

When you brake, you squeeze the brake lever against handlebars. The brake translates this force onto pads that squeeze the rim/rotor. If you imagine squeezing the front wheel with your hand (don't try this in practice!), you realize that you wouldn't get much brake power. Your hand isn't strong enough to squeeze the rim hard enough to slow the bike.

The mechanical advantage of the brake mechanism amplifies your hand power to squeeze the rim harder than you could with your hand. This is easiest to visualize by looking at a set of pliers: The jaws represent the lower brake arms that squeeze the rim. The handles – the upper arms – are longer, giving your hands more power. However, the lower arms (jaws) will not move as far as the upper arms (handles).

On a bicycle brake, the brake pads move a much smaller distance than the brake lever. For example, you squeeze the brake lever by 25 mm (1"), but the brake pads only move 5 mm (0.2"). The mechanical advantage is 5, and the pads clamp the rim with five times the force you used to squeeze the brake lever.

More mechanical advantage reduces the hand force on the brake lever, but it also reduces the distance the brake pads travel. On brakes with high mechanical advantage, the brake pads must be set very close to the rim/rotor, otherwise the brake lever will bottom out against the handlebars during hard braking. And the brake pads will rub if the rim/rotor goes out of true or the brake goes out of adjustment.

The pads on a brake with low mechanical advantage can be set far from the rim/rotor, but the rider must squeeze the

A caliper brake's mechanical advantage is determined by the ratio between the upper and lower arms.

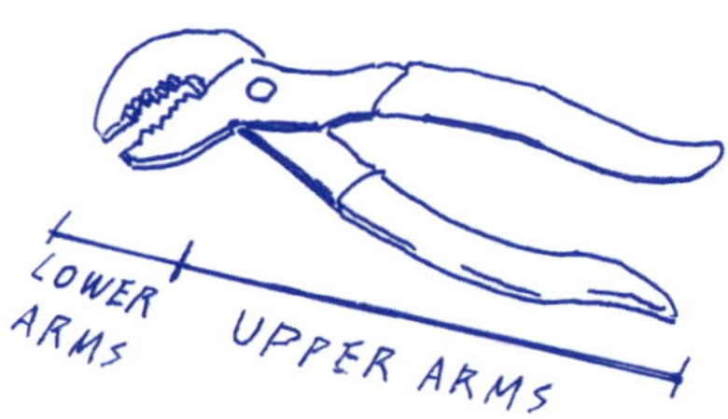

brake lever with more force to obtain the same braking power. If the rider does not have enough hand power, this will limit the absolute braking power of the bike.

Both brake lever and brake caliper amplify the rider's hand force. Their mechanical advantages are added to obtain the mechanical advantage of the system. Brake lever and brake must be matched to provide the desired mechanical advantage.

On brake levers, the mechanical advantage depends on the location of the pivot in relation to the lever. Traditional brake levers with the cable exiting at the top usually have lower mechanical advantage than modern 'aero' brake levers. That means they pull more cable for a given lever travel.

On the brake, the mechanical advantage is determined by the ratio of two levers: the lower lever between pivot and brake pad, and the upper lever between pivot and cable anchor. Centerpull and cantilever brakes have straddle cables that also can influence the brake's mechanical advantage.

When you adjust a rim brake pad at the bottom of the slot, you decrease the brake's mechanical advantage, and vice versa. This can be an issue especially with brakes that have their pivots close to the brake pads. Some models offer 50% more mechanical advantage with the pads at the top than with the pads at the bottom of the slot.

Another consideration is the size of the wheel/rotor: Smaller rotors have less mechanical advantage – whether they are rotors with disc brakes or the bicycle's rim with rim brakes.

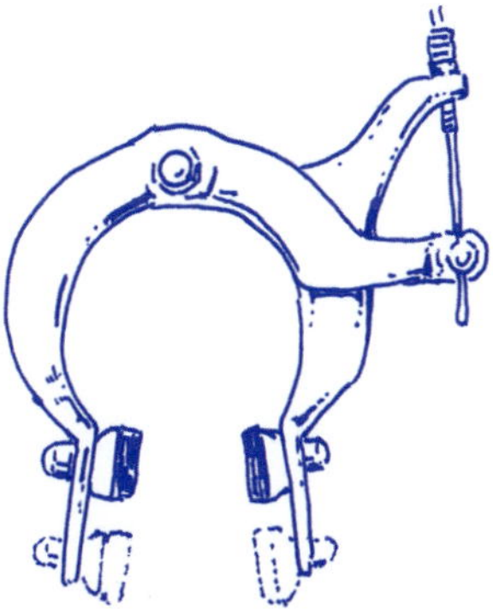

Moving the brake pads up or down in the slots of the brake arms changes the mechanical advantage.

Friction Losses

Not all your hand power is transmitted to the brake pads. Most of the losses are due to cable friction. These losses can be significant. Cable friction increases if the cable is routed in tight bends. This can occur with modern handlebars that have tight curves, or when the brake cable is routed under the stem into a cable stop. Route brake cables in the widest curves possible. Hydraulic brakes use fluid to transmit the braking power with almost no losses.

Flex

If a brake is too flexible, it can feel fine at first, but as you squeeze the lever harder, the brake force does not increase linearly. Instead of squeezing the rim or rotor harder, increasing hand force just flexes the components of the brake system.

Flex occurs primarily in the caliper.[51] Brakes with long arms to fit over wide tires can be very flexible. The flex occurs mostly where the pads are pulled along by the rim as you brake. This means that only the lower arms that connect the pads to the pivots flex significantly. The upper arms, between pivots and cable attachments, are stressed only in one plane, so it's easy to make them rigid without adding much weight. The best solution for reducing brake flex is to make the lower arms as short as possible by locating the pivots near the brake pads. There is also flex in the mounting system of the calipers. Direct-mount brakes, which attach to the fork blades and chainstays, are more powerful than brakes that attach to the fork crown.

Flex also occurs in the brake cable housing. It compresses as you squeeze the brakes. 'Compression-less' housing can help, but the best solution is to keep the housing short. That is why most modern bikes don't run continuous brake cable housing to the rear brake, but instead have housing stops on the top tube.

Flex also occurs in ferrules or others part that move slightly as you squeeze the brake. Cable hangers for cantilever brakes, inline adjusters and secondary 'safety' brake levers that are spliced into the cable housing are notorious for adding flex.

To test brake flex, squeeze the brake lever hard and watch the front brake. Flex should only be visible in the brake pads as the rubber compresses, but not in the brake itself.

Modulation

Modulation is almost as important as absolute brake power. On paved descents, you can brake deep into turns, so the weight transfer increases traction on the front wheel. Approaching a turn during a fast gravel descent, you want to brake as hard as possible without locking up the front wheel. In these situations, the brake should react in a linear fashion as you squeeze the brake lever. Brakes with good modulation allow you to meter out the braking power with precision.

A brake with poor modulation can lock onto the rim and increase its braking power suddenly, even though the rider has not changed the brake force. This is usually caused by flex in the brake arms or their attachments, which changes the angle of the pads on the rim. Stiff brakes usually offer good modulation. On early disc brakes, brake power often increased suddenly as the pads got hot in mid-corner. Advances in pad design have eliminated this problem for the most part.

Racing Bikes

Modern rim brakes for racing bikes are very powerful and offer excellent modulation. With their short arms – they only need to reach around a narrow tire – modern road brakes are stiff and don't lose much power to flex. Direct-mounting the brakes to pivots on the fork blades reduces weight and improves performance further.

Wide Tires

To reach around wide tires, caliper brakes need to be long, which makes them flexible and/or heavy if they are beefed up to reduce flex. Direct-mount brakes – cantilevers or centerpulls – eliminate most of this problem by moving the pivot close to the rim to limit flex.

Disc brakes are popular for bikes with wide tires, because they eliminate the need for the brake caliper to reach around the tire. It only has to enclose a thin rotor, so the caliper can be very stiff, and, in theory, offer superior braking.

Wet Conditions

When it rains, the braking surfaces of rim or disc rotor get covered with water. Before the brake can work, the pads have to wipe off the water. A brake with higher mechanical advantage squeezes the rim/rotor harder and wipes off the water faster. Disc brakes have a high mechanical advantage (to make up for their small rotors), so they tend to work well in the rain. Even inexpensive disc brakes stop well in wet conditions.

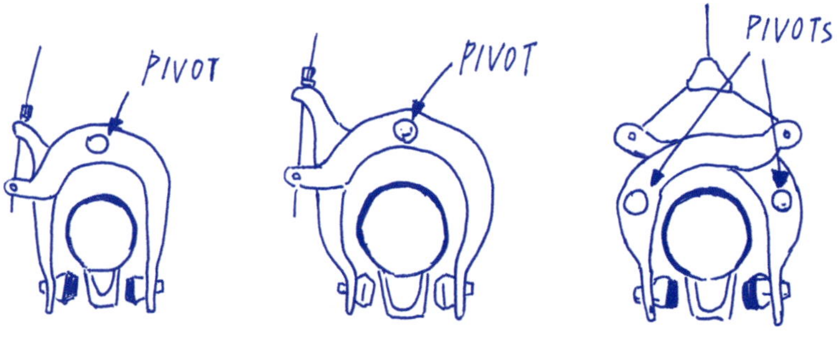

A classic rim brake (mounted to the fork crown) works fine for narrow tires (left). To span around a wide tire requires a much larger caliper (middle), which increases flex and weight. Centerpull brakes (right) move the pivots closer to the rim to limit flex.

Poorly designed rim brakes may offer adequate braking in dry conditions. Many riders never brake hard enough to notice the flex that limits their ultimate braking power. However, in the wet, some brakes are unable to wipe the water off the rim. No matter how hard the rider pulls on the brake lever, the bike will not slow down quickly. This dangerous situation is not uncommon with budget long-reach brakes that are sold for all-road bikes with large tires.

With rim brakes, once the rim is dry, braking will generate heat so rainwater hitting the rims will evaporate quickly. During long wet descents, riders can drag the brake slightly to heat the rim continuously and keep the brake ready to operate.

Disc Brakes

Disc brakes use a separate rotor, rather than clamping the wheel rim with their brake pads. Disc brakes have three main advantages:

• The calipers don't need to reach around the tire, so they can be very stiff, with just a slot into which the rotor fits.

• Because the rotor diameter is small (compared to a rim), disc brakes must have an extremely high clamping force. This helps wipe off water, so discs perform better in the rain than all but the best rim brakes.

• Disc brakes do not heat up the rim, so there is no risk of melting the tire during prolonged braking, which can cause blow-offs. (With rim brakes, proper technique – braking hard for each turn and letting the bike roll in between – eliminates this risk on all but the steepest terrain.)

On the downside, disc brakes tend to be heavier than rim brakes. The calipers have to be very strong to clamp the rotor hard, so they are heavier than rim brake calipers. Add the weight of the rotor, and the weight penalty can be significant. However, disc-specific rims don't need braking tracks, so carbon rims work well with discs. The weight savings of carbon rims can almost make up for the extra weight of the separate rotors and heavier calipers.

Disc brakes require stiffer fork blades, since the reaction to the braking force occurs low on the fork, rather than near the fork crown (as on rim brakes). If the fork blades aren't stiff enough, the left fork blade will flex backward, affecting the bike's steering as the rider brakes. This is the main reason why many riders prefer rim brakes: They allow the use of relatively flexible fork blades that offer suspension to improve comfort and speed. Mountain bikes have separate suspension, combining stiff fork blades with shock absorption. That eliminates the main disadvantage of disc brakes – virtually all mountain bikes today are equipped with disc brakes.

With disc brakes, it is extremely important that the hub is securely mounted in the fork, otherwise, the front wheel can eject during braking. Thru-axles have eliminated this problem, but forks with traditional dropouts require 'lawyer lips' to prevent accidents.

With disc brakes, the left fork blade is pushed back when braking hard, and the front wheel tries to pull out of the bike's dropouts. Stiff fork blades solve the former problem, thru-axles the latter.

Mechanical vs. Hydraulic Discs

Because the rotor is relatively small compared to the diameter of the bike's wheel, the mechanical advantage of disc brakes must be very high. This magnifies any flex in the system, which will rob valuable braking power. Mechanical disc brakes should be set up with compression-less housing. Careful installation of the brake cables is key.

Most mechanical disc brake calipers use a spiral track on which a tiny steel ball runs. The track rotates when the brake is applied, and the ball is pushed outward, which in turn pushes the pads together. This mechanism has significant friction and reduces the power of the brake. Mechanical disc brakes with rotors that measure 160 mm or smaller rarely offer the same stopping power as a good rim brake. Larger rotors could solve this issue, but they aren't usually spec'd on all-road bikes.

The best hydraulic disc brakes offer outstanding stopping power and excellent modulation. Hydraulic brake lines eliminate the flex in the brake cable and housing. Hydraulic calipers use pistons that clamp the rotor directly, with almost no friction losses. Even though their theoretical mechanical advantage is the same as that of mechanical disc brakes, they tend to be far more powerful. Some models can feel grabby at low speeds, but most current hydraulic disc brakes offer excellent performance.

A disadvantage of hydraulic brakes is the more difficult setup, which exceeds the capabilities of many home mechanics. Hydraulic brakes also require more frequent maintenance and rebuilds than cable-actuated brakes.

Brakes Summary

Differences in absolute brake power are most apparent in the wet, when poorly performing brakes take a long time to wipe water off the rims. Until the rims are dry, the rider experiences almost zero braking power.

For racing bikes with relatively narrow tires, modern rim brakes offer great stopping power and modulation. Rim brakes for wide tires must be carefully designed to limit flex. The best centerpull and cantilever brakes remain a viable, ultralight option. Most disc brakes bite well in the wet, but many mechanical discs still lack absolute stopping power. Hydraulic discs are very powerful, but not always easy to service at home. All disc brakes require stiffer forks to the detriment of comfort and speed.

Brakes on Motorcycles and Cars

Why do virtually all motorcycles and cars use disc brakes, yet for bicycles, rim brakes remain a viable alternative? Cars and motorbikes cannot use rim brakes: Their tires are too big, and brake calipers cannot reach around them without becoming so big and bulky that they are impractical.

Before disc brakes, cars and motorbikes used drum brakes: brake shoes push against the inside of a spinning drum to slow down the wheel. Drum brakes can offer excellent braking, but they suffer from overheating, since the braking surface is on the inside of the drum. That is why disc brakes have become popular for cars and motorcycles.

One disadvantage of disc brakes is that they work poorly with cable operation, so cars and motorcycles use only hydraulic brakes. However, a car's parking brake is always controlled by cables, so it remains engaged no matter how long the car is parked. Most cars use small drums inside the rear discs for their parking brakes, rather than try to activate the discs with cables.

This parallels the developments in the bicycle world: Drum brakes never became popular on bicycles because they tend to fade as they get hot. Cable-actuated mechanical disc brakes suffer from poor performance, but hydraulic discs can offer extremely powerful braking.

Drivetrain

"A slight tug on the delicate-looking lever produces an instantaneous shift. No overshifting is required. Shifting in mid-hill is easy. Gears engage securely, and once selected, they remain engaged. The constant chain tension minimizes chainslap on the small chainring."

More than 130 years after the safety bicycle with two equal-sized wheels was invented, most bikes still use a chain to transmit the power from the pedals to the rear wheel. On cars, chain drives were replaced by shafts almost a century ago, but the 'old-fashioned' chain persists on bicycles – because it remains the best solution.

Shaft drives on bikes actually predated their use in cars. Shaft drive bicycles were popular a century ago, but the need for precise manufacture and high frictional losses made them disappear again. Toothed drive belts, which in car engines have replaced chains for powering overhead camshafts, have become popular on bicycles in recent years. Belts don't require lubrication – resulting in a cleaner drivetrain – and they make less noise. This is offset by the need to provide a way to open the frame's rear triangle, since the belt cannot be split during installation like a chain. And since belts are stiff, they must be perfectly aligned, so they cannot be used with derailleurs. That limits them to bikes with single-speeds or gear hubs.

The concept that underlies derailleurs makes engineers shudder: To change gears, you bend the chain until it jumps onto the next cog. Depending on your gear, the chain may not run straight. A spring takes up the chain slack. It seems crude, and yet it remains the most efficient system.

Before multiple gears, even Tour de France champions walked up mountain passes.

Gear Hubs vs. Derailleur

Gear hubs predate derailleurs by a few years, and they remain popular in some applications. Their fully enclosed mechanism is impervious to dirt. The chain runs on only one set of cogs, so differential cog wear isn't a problem as it is on derailleur systems. Both these factors combine to make gear hubs almost maintenance-free. Well-designed multi-speed hubs have evenly-spaced gears without duplication and with a simple shifting sequence.

Why don't we all use gear hubs? Like shaft drives, gear hubs are less efficient because of all the cogs that mesh and turn. The more gears a hub has, the higher its resistance. The best gear hubs use ball or roller bearings, which reduce the resistance a bit. Apart from one gear that uses direct drive, the efficiency of gear hubs decreases sharply as the rider's power output increases, whereas the resistance of derailleurs is almost constant, independent of power output.[52]

These factors – no maintenance, but lower efficiency – make gear hubs most suitable for city bikes and a more leisurely riding style, while derailleurs remain the choice for most performance bikes.

There is another difference that is often overlooked: The extra (rotating) weight of a gear hub changes the feel of the bike. Lifting the rear wheel over bumps becomes more difficult, and especially with larger hubs that have more gears, rocking the bike from side to side when riding out of the saddle requires more muscle. A bike with a gear hub simply feels different from a lightweight performance bike. On the other hand, if you prefer to ride seated with moderate power outputs, a quality gear hub can be an excellent choice. A test ride will show you whether you like it!

Gearing

Derailleur systems have changed in recent years, too. It used to be that racers used doubles, cyclotourists triples, and single chainrings were only found on cheap bikes for casual use. Today's rear cassettes have so many cogs that two chainrings are enough for most riders. Triples still have a place for riders who need a very wide gear range. And many mountain and all-road bikes now use a single chainring with a wide-range cassette on the rear to eliminate the rhythm-breaking and potentially troublesome front shifts.

Gone are the days when most road bikes came with huge chainrings – 53 and 39 teeth were common. Professional racers need these monster gears when they sprint toward a downhill finish with a strong leadout. For the rest of us, chainrings that large mean we rarely use the lower half of the rear cassette. Much better to downsize on the front and use a cassette with closer ratios on the rear to keep the steps between gears at less than 10%, so our spin isn't interrupted every time we shift.

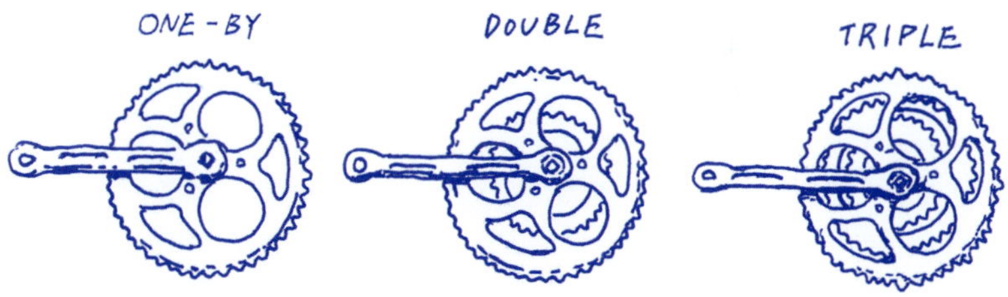

One-by drivetrains simplify the shifting. Double chainrings allow smaller steps on the rear. Triples are useful for riders who need a wide spread of gears.

To determine the gearing you need on your bike, start by thinking about these three gears:

- High gear: The largest gear that you use when you are sprinting for a city limit sign with friends, or when you are riding with a powerful tailwind. You won't need a larger gear than this.
- Base gear: The gear you use when spinning along on flat roads with no wind. This gear should be in the middle of the rear cassette, so you can adjust to changes in speed and terrain with a simple shift or two in the rear.
- Low gear: The smallest gear you need on the roads you usually ride.

The high and low gears are easy to determine, but the base gear is actually the most important. It depends both on your strength and your preferred cadence.

Double or Triple?

The decision comes down to the gear range you need and which gears you ride most of the time. If your base gear is close to your high gear, then you can accommodate both on the same chainring. That means that you can use a double. Modern cassettes start with 11- or 12-tooth cogs, which means that a 46-tooth big ring is large enough for most riders. More than 16 teeth difference between chainrings can result in poor shifting performance, so a 30-tooth small ring is a good choice. A 46x30 with a 12-30 cassette will work well for many riders.

If your base gear is much lower than your high gear, then you'll be less happy with a double. Too often, you'll ride on big ring and the largest cog on the rear. Even the slightest uphill will require a shift to the small ring. As the road flattens, you'll need to shift to the big ring again. All those front shifts require rear shifts to compensate for the big step – you'll be shifting most of the time! Better to get a triple, with a middle ring that puts your base gear near the middle of the cassette. For example, a rider who doesn't spin at a fast cadence (and thus needs a bigger high gear) might do well with a 48x42x28 and a 11-32 cassette.

This chainring combination may look a bit different from most triples that space the chainrings more evenly – like a 48x38x28 – but it's natural for most riders to have their base gear closer to their high gear than to their low gear. For most riders, a 38-tooth ring is too small for normal riding on the flats.

Triples have some disadvantages: They don't always shift as well as doubles, and their Q factor is wider. Many riders can use a double if they are willing to reduce their high gear (and perhaps work a little on their spin). A 44x28 would offer better shifting and a narrower Q factor than the 48x38x28 mentioned above.

One-By

Many mountain, cyclocross and gravel bikes use a single chainring. Using a single shift lever and having the next gear always just one click away removes the need to think about shifts. A 'one-by' drivetrain provides the convenience usually associated with gear hubs. Eliminating the potentially troublesome front shifts – including the potential for chain suck – provides an additional plus, especially on rough terrain.

Rear derailleur geometry can be much simpler when the front chainring size is constant. This further improves shifting performance. Eliminating the second chainring, front derailleur, shifter, as well as associated cable and housing saves weight, although larger rear cassette cogs offset some of the savings. To keep the chain from derailing off the single front chainring when it is running at extreme angles on the largest and smallest rear cogs, one-by chainrings have special teeth that mesh perfectly with the chain's inner and outer links.

Downsides of one-by drivetrains are a poor chainline in the extreme gears and relatively large steps between gears. One-by drivetrains work best for strong riders who don't need a very wide gear range, or for riders who can put out power over a wide range of cadences that make the large steps between gears a non-issue.

Electronic vs. Mechanical Shifting

Mechanical shifting uses cables to pull the derailleurs into position. Electronic shifters use servo motors in the derailleurs, which are activated by buttons that are integrated into the brake levers. The derailleur geometry is the same for both systems, so the actual shift is no different – in theory.

In practice, the motors of electronic derailleurs push the chain with great force, which results in positive shifts every time. On mechanical systems, the ratchets inside the shift lever can wear out quickly, especially if the rider prefers rapid shifts and releases the lever before it engages completely. Shifter cables can get contaminated with dust. Electronic shifting tends to be more reliable and trouble-free than mechanical brake/shift levers, but it requires keeping the batteries charged.

Downtube and Bar-End Shifters

Downtube shift levers were superseded by more modern shifters more than 20 years ago, yet many riders still prefer them. They are lighter, not only because the levers themselves weigh less, but also because they do not require long (and heavy) cables and housing to the brake levers. Down tube levers can be moved with the entire hand, making them easier to operate with cold or tired hands than other systems.

For riders with a relatively inclined position, downtube shifters are at the same distance as the handlebars, making them easy to reach.

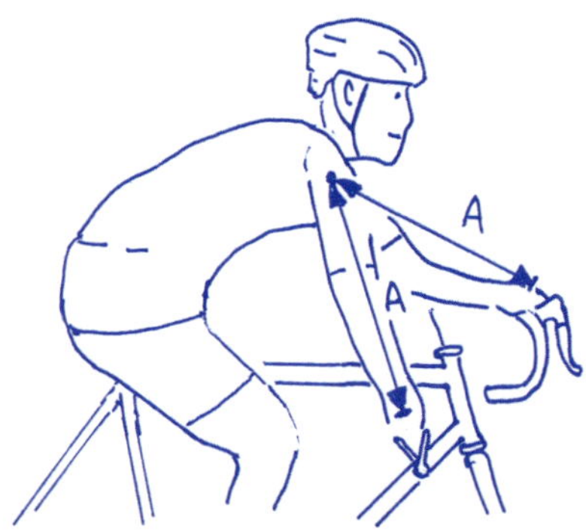

For many riders, downtube shifters are very ergonomic: On a performance bike with drop handlebars, downtube shift levers are equidistant from all hand positions. This makes the shift levers always easy to reach, for example, when climbing with the hands on the tops of the handlebars, whereas modern brake/shift levers require a long reach to execute a shift.

Bar-end shifters offer most of the simplicity of downtube levers without the need to remove a hand from the handlebars during shifts. This can be useful on bumpy terrain and explains why bar-end shifters were popular in cyclocross before integrated brake/shift levers replaced them.

Cranks

Modern cranks integrate the spindle, which has thin walls and a large diameter to save weight. This works especially well with carbon cranks, which have to be large to make up for the lower strength-per-volume of composite materials.

Oversized spindles leave less space for the bearings. Some makers resolve this by moving the bearings to the outside of the bottom bracket shell, but bearing alignment can be an issue. Fortunately, most systems make it easy to replace the bearings.

Square-taper bottom brackets still have many followers. This design allows for larger bearings, and the crank/spindle interface is self-healing and reshapes itself every time the crank is installed. The spindle length can be selected freely to fine-tune your chainline or make room for extra-wide chainstays.

Cranks also have changed as manufacturers finally have realized that most cyclists are not professional racers and need smaller gears. Compact cranks allow for smaller chainrings, and subcompact cranks are becoming popular.

Probably the biggest factor affecting crank performance is the Q factor (also called tread) – the horizontal distance between the pedal eyes. This determines your stance as you pedal. Many riders are able to put out more power with a narrow Q factor (see p. 70), yet cranks have become wider in recent decades. Most racing cranks have a Q factor of 146 - 150 mm, which seems to be the widest that many performance riders can easily tolerate. Triple cranks obviously are wider, and so are mountain bike cranks that must clear ultra-wide tires and beefy chainstays. This reduces many riders' power output. It can also make it harder to achieve the 'effortless spin' that makes cycling on a good performance bike so much fun.

Crank length is a topic of much discussion. It makes sense that taller riders should have longer cranks, but common crank lengths vary only between 165 and 175 mm. The 10 mm difference amounts to only 6%, when riders' leg lengths easily vary by 20% or more. Some feel that the small differences between most cranks are almost too small to detect. Others believe that tall riders should use much longer cranks, and short riders much shorter ones. That means, of course, that they also would have to spin their legs at very different cadences, otherwise, taller riders would be 20% faster than shorter ones!

Some tall riders struggle to spin at 100 rpm or more, and they tend to bottom out between pedal strokes – which stresses their bike and reduces their power output. These riders may want to try longer cranks than the commonly available sizes. Short riders who find standard-sized cranks difficult to spin may benefit from ultra-short cranks.

Cyclists' body dimensions vary by 20% or more, yet most crank lengths cover only a range of about 6%.

Chainline

The chainline determines where the chain runs in relation to the centerline of the frame. It is measured on the middle chainring (or between the chainrings on double cranks) and on the middle cog on the rear. For decades, road bikes had a chainline that was 43.5 mm from the centerline of the bike.

In recent years, the front chainline has moved outward to make room for wide tires and large chainstays. Many cranks today have a chainline of 45 mm, and mountain bikes use values between 47 and 50 mm. On the rear, the chainline has moved inward as makers have added cogs to the cassette. The chainline between the chainrings at the front and the cassette at the rear should match within a few millimeters, otherwise, there will be excessive cross-chaining in some gears. This is one reason why mixing mountain and road bike components can result in sub-optimal results: The cogs in the rear may not line up properly with the chainrings at the front.

Pedals

Almost every rider swears by their favorite pedal system. Walkable shoes are a plus for anybody who isn't racing. It's much nicer to walk into a cafe than to hobble. On the bike, the power transfer of most pedals is similar. More important is that the shoes are stiff enough to prevent pressure points – which can lead to nerve damage on the balls of your feet – yet flexible enough to prevent discomfort during long days in the saddle.

Classic pedals with toeclips and straps still have their fans. They are light and can be used with a variety of shoes. The toestraps can reduce circulation in your feet, which can make them numb during long rides and cold during the winter.

During steady efforts, pedals without foot retention do not give up much in efficency and speed. Most riders only pull up on the pedals during short bursts of acceleration, when foot retention allows them to put out more power.

Modern vs. Classic Standards

"The bike is handicapped by its massive T47 bottom bracket shell and 44 mm tapered head tube. These parts are dimensioned for carbon, and replicating them in steel adds significant weight."

Carbon fiber is fundamentally different from metals. Metals require relatively little material to make strong frames and parts. Carbon fiber is much less dense, so it requires large volumes to withstand the rigors of hard riding. Metal works fine with angular shapes, as long as stress risers are avoided, while carbon requires an uninterrupted flow for strength.

As the preferred frame material has changed from steel to carbon, so have the standards for bicycle components. The new standards work well with carbon frames and forks, but they are less suitable for metals, especially steel. Attempting to get the 'best of both worlds' by combining classic materials – steel frames and aluminum components – with modern standards can result in bikes that are heavier and less performing than the classic bikes they are trying to improve.

'Standard-diameter' steel tubing became standard because it offered the best ride and performance for strong racers. When titanium and aluminum frames became popular, their lower stiffness required oversized tubing. Trying to remain competitive with the new materials, the makers of steel tubing introduced thinner-wall tubing, in a larger 'oversize' diameter to maintain the same stiffness as on the 'standard' frames that worked so well.

The new steel bikes were a little lighter, they looked a little more like the popular titanium and aluminum bikes, and their excellent ride characteristics were mostly unchanged. Things started to go awry when the new oversized tubing diameter was applied to lower-end frames – without reducing the wall thickness. These frames may have looked 'modern,' but they also became stiffer than was ideal for most riders.

When carbon fiber became the preferred frame material, its lower strength-to-volume ratio required an almost complete redesign of the entire bicycle. Steerer tubes on steel forks for single bikes measured 1" (25.4 mm) in diameter. (28.6 mm for threadless steerers that could be made with thinner walls and larger in diameter, since they didn't need to accommodate

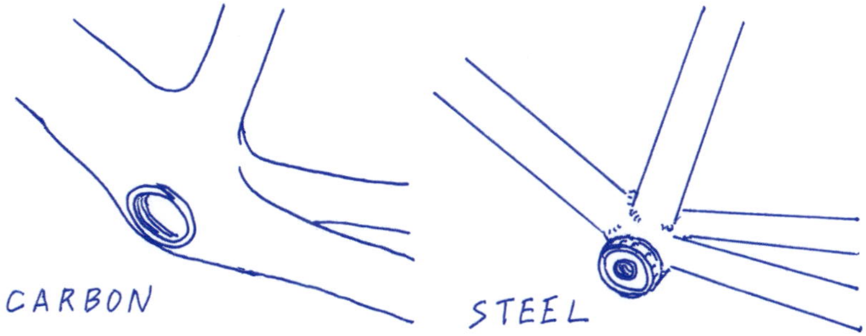

Carbon bikes tend to be bulky, steel bikes are more slender, because of the properties of their materials.

headset threads.) This diameter offered the right strength and stiffness. Tandem forks are subject to much higher loads, and the best tandems used oversized, tapered steerers. This technology was not applied to single bikes because it offered no advantages as long as the forks were made from steel.

When the first carbon forks were made to the 1-inch standard, they were not strong enough and developed a reputation for sudden failures. The solution: The diameter of the steerer was increased to make it stronger. Most carbon forks now have tapered steerer tubes – basically replicating old tandem forks in carbon to make up for carbon's lower strength-to-volume ratio. Building a steel frame and fork to modern carbon standards makes them heavier than necessary. In effect, it is putting a tandem fork on a single bike.

Carbon frames have also led to a redesign of the components. Square taper spindles were dimensioned to keep the (heavy) bottom bracket shells of steel frames as small as possible. The small spindles also allowed for slender, lightweight cranks with just enough aluminum to surround the spindle. Carbon frames require much more material volume for strength and stiffness, especially in the bottom bracket area. Installing a slim square-taper bottom bracket makes little sense, as there is room for a much-larger, thinner-walled spindle that can save some weight. If the cranks are also made from carbon, they need to be bulky as well, and they'll have plenty of space to accommodate a large spindle.

Using the same 'carbon standards' for aluminum cranks and steel frames increases the bike's weight, because both the crank and the bottom bracket shell need to be bulkier than would be necessary otherwise, to surround the larger spindle. Large, thinwall spindles save weight on a carbon bike, because they use the available space more efficiently, but they add weight to a steel bike because they require more material to accommodate them – which negates the weight savings of the lighter spindle itself.

Disc brakes do not work well with dropouts and quick releases, because the wheel can move or even eject during hard braking (see p. 190). Thru-axles have solved this problem. The standards for 12 mm thru-axles on all-road bikes were designed for carbon forks, where the fork blades are bulky all the way to the tips. On steel forks, massive fork ends are necessary to surround 12 mm holes, and that adds weight.

The effects of these new standards cascade through the bike. The oversize steerer tubes of carbon forks mean that the stems also need to be larger. To make the handlebars fit seamlessly, their diameter has to increase as well – which is required anyhow if the bars are made from carbon. (In this case, making aluminum handlebars with a larger center bulge adds relatively little weight, so riders of carbon bikes can benefit from the greater impact strength of aluminum bars.)

The new standards make sense for carbon bikes, because they were designed for these bikes, just like the old standards made sense for steel bikes. The old standards have become obsolete not because the old builders and component makers did not know what they were doing, but because different materials require different dimensions.

If the goal is to optimize performance and minimize weight, it makes sense to build the bike with the standards that were developed for its material. Since modern components are designed for carbon bikes, this means that steel bikes are falling behind, not because of some inherent deficiencies of steel as a frame material, but because most components are no longer optimized for steel's requirements.

Steel bikes still have a place if they capitalize on the strength of their material: Steel's density allows it to be slender and flexible. Steel fork blades can absorb shock better, and steel frames can be tuned to flex and 'plane.' As an added plus, steel bikes are easy to customize. To realize the full benefits of steel, steel bikes should be built to 'steel standards,' and not replicate carbon bikes in steel.

Contact Points

"One thousand kilometers into Paris-Brest-Paris, my body is free of aches and pains. Only my hands feel the relentless vibrations of the handlebars, even though my bike is equipped with 42 mm-wide tires."

The contact points are where your body touches the bike: saddle, handlebars and pedals. They determine the first and most important layer of comfort on your bike: Even the most supple tires and shock-absorbing fork blades will not make your ride comfortable if your hands or bottom hurt because the shape of your handlebars or saddle does not match your anatomy. (Pedals were discussed in the previous chapter.)

Handlebars

Handlebars are often overlooked, yet they are an important element for comfort and injury prevention. During the 'heroic age' of the 1920s and 30s, when stages were long and roads were rough, racers had their handlebars custom-bent to their perferred shapes, and transferred their favorite bars from one bike to the next.

During a long ride, you hold the handlebars for many hours, while they vibrate with road buzz. A poorly shaped handlebar can cause nerve damage and result in tingling or numb fingers, hands, and arms. Drop handlebars offer multiple hand positions. This allows changing your riding position to accommodate variations in speed, and it also helps alleviate pressure on your hands.

Many modern handlebars have relatively square shapes with tight-radius curves. They offer only three distinct positions: brake lever hoods, tops, and drops. Yet most riders' wrists don't fall naturally at 90° angles. Classic bar shapes with generous curves allow riders to find intermediate positions on the ramps that match the natural angle of their wrists.

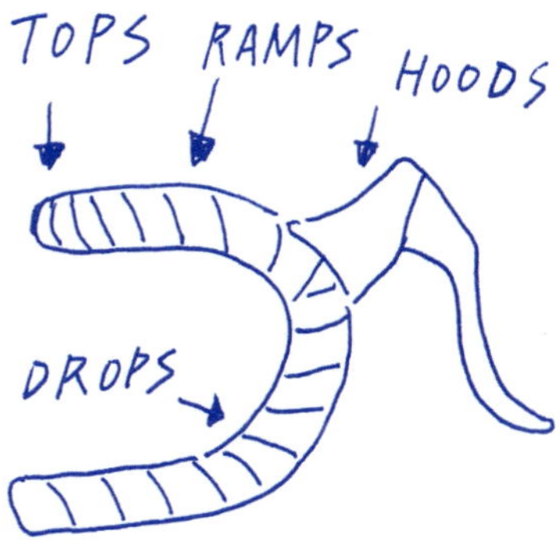

Drop handlebars provide multiple hand positions.

Reach

Handlebar reach is measured from the 'tops' to the most forward part of the 'drops.' The reach determines how much room there is in the 'on the ramps' position for your hands. In this position, the rider holds onto the handlebars behind the brake levers.

Generally, handlebars with longer reach have more room for the rider to roam and find a comfortable hand position. Short-reach handlebars lock riders into a few hand positions that may not match their anatomy. If you install handlebars with longer reach, you'll need a shorter stem – but not by quite as much as you may think. You'll want to split the difference in half: If your bars are 20 mm longer, install a 10 mm shorter stem. Drop bars offer multiple hand positions, and you'll probably ride most of the time in an intermediate position between the hoods and the tops.

Drop

Handlebar drop is the measurement between the ramps and the drops, measured center-to-center. Handlebar drop, together with handlebar reach (see above), determines how

Reach, drop and width determine the dimensions of a handlebar.

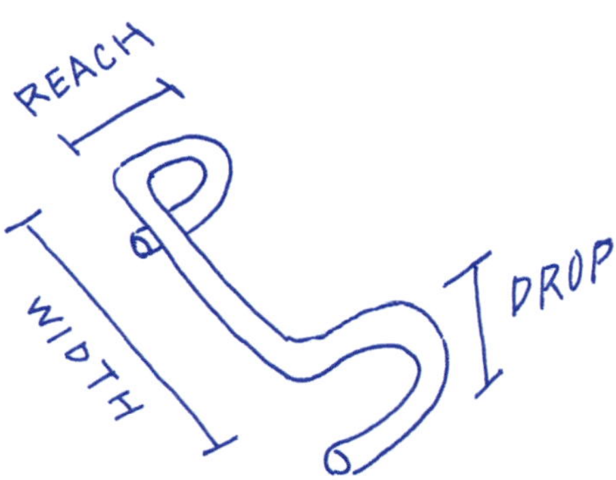

much difference there is between the various riding positions that the handlebars offer. A deep-drop handlebar has distinctly different positions 'in the drops' and 'on the hoods,' whereas with shallow-drop handlebars, the angle of the rider's back is almost the same in both positions.

If the rider desires both an upright position 'on the hoods' and a performance-oriented 'in the drops' position, then a deep-drop handlebar can be combined with a higher stem to offer both positions. However, if the drop is too deep, then the two positions may be too far apart, leaving the rider searching for the elusive 'in-between' position.

Width

Many riders believe that handlebar width should match their shoulder width, because otherwise their breathing will be restricted. However, our arms articulate if we bend our elbows (which is good form for many reasons), and we can breathe fine even when we climb with our hands 'on the tops.' Thus handlebar width does not affect our ability to breathe.

If the rider bends their elbows, narrow handlebars do not cause shoulder pain or restrict breathing.

The width of the handlebars determines how much leverage we have over the steering. This means that bikes with high-trail geometries (and more wheel flop, see p. 80) usually have wider handlebars, whereas low-trail geometries that only require a light touch on the handlebars work best with narrower bars. Handlebars that are too wide can cause shoulder pain, whereas handlebars that are too narrow can make it difficult to control the bike.

Often, riders who bend their elbows prefer narrower handlebars, because wide bars force their hands outward when they naturally articulate inward. Conversely, riders who lock their elbows are more comfortable on wider handlebars, because narrow bars constrict their shoulders.

Shape

Reach, drop and width only express the basic fit of handlebars. The shape – the radius of the curves, the angle between ramps and drops, etc. – is just as important for comfort. Most riders find generous curves more comfortable, especially when riding long distances.

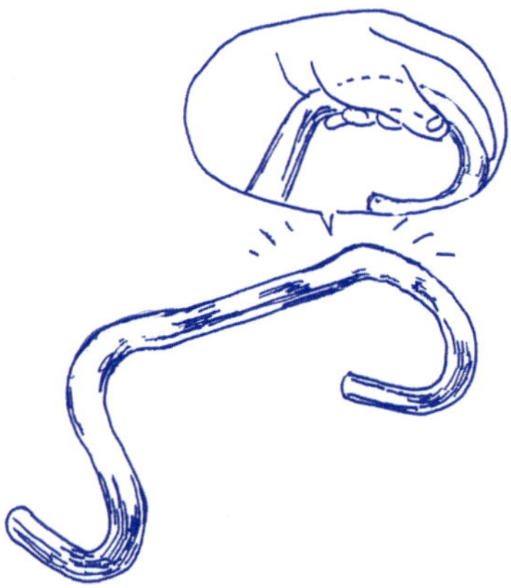

The upward sweep of randonneur handlebars creates a convex shape that supports the concave palms.

Randonneur handlebars sweep upward next to the stem. This provides a three-dimensional curve on the ramps that matches the cup of the rider's hands. When randonneur handlebars are shaped just right, they can provide supreme comfort even during the longest rides.

Saddle

Your riding position as well as your anatomy will determine the choice of your saddle. Strong riders with a low position lean forward and have little weight on their sit bones, so relatively narrow saddles work best for them. The upward thrust from pushing down on the pedals reduces the weight that rests on the saddle, so powerful riders are less likely to suffer from saddle problems. A more upright position places the rider's sit bones more firmly on the saddle, while the lower pedaling force means that more weight rests on the saddle. This usually requires a wider (and often softer) saddle to support the weight.

The actual shape of your saddle is a very personal choice. Most riders eventually find a model they like. It used to be common among experienced riders to transfer their favorite saddle from one bike to another.

Anatomic saddles with grooves and cutouts can improve comfort for some riders. Others find that the cutouts result in increased pressure from the remaining parts of the saddle.

Classic leather saddles shape themselves to the rider's anatomy. They require some breaking in. After that, many riders find them supremely comfortable. Where modern 'anatomic' designs can only approximate the rider's bottoms, classic leather saddles shape themselves over time to provide a perfect fit.

Try different saddles until you find one you like, and don't worry about its weight. Comfort is more important for performance than a few grams.

Racks and Luggage

"The tires of our bikes hum on the pavement, a slight tailwind pushes us along, and life could not be better. Even though our bikes are loaded with panniers, they still feel like performance bikes – because that is what they are."

Bike racers have support cars that carry spare clothes and food, but for the rest of us, the ability to carry some luggage is useful even on day rides. We need a place for the long-sleeve jersey we take off as the day warms up; we want to carry a picnic lunch for a ride into the mountains far from the next town; and we like to bring a camera to document our adventures. Fortunately, bikes can carry all kinds of loads and still be fun to ride.

Front vs. Rear Load

Traditionally, bikes have often carried their loads on the rear. It's easy to attach a rack to the rear triangle, and placing it there doesn't seem to greatly change the handling. But the rear is not the best place to carry a load. Here is why: A bike is balanced by moving the front wheel from side to side. The front of the bike is balanced directly by turning the handlebars. The rear cannot be steered directly. It follows the front, but with a lag and with reduced amplitude. (The rear tire runs on a straighter path than the front.)[53]

To balance a rear load, you move the rear tire from side to side by first moving the front tire and then waiting for the rear to follow. This requires larger inputs from the rider than balancing a front load, and there is a longer lag until the inputs have an effect on the load.

All-road bikes can carry a camping load without losing their spirited feel.

Furthermore, a rear load is unsupported when the rider rises out of the saddle. The load becomes the proverbial 'tail wagging the dog,' making it difficult to ride out of the saddle. A front load is attached to the fork and thus connected directly to the handlebars. When riding out of the saddle, the rider always holds onto the handlebars, so riding out of the saddle with a front load feels little different than it does without a load.

A load carried on a rear rack requires a bike with a stiff frame to avoid flex that negatively affects the handling. Stiffer frames often don't perform as well as more flexible ones (see p. 60). A front load rests almost directly on the front wheel, so it can be carried on a high-performance frame without modifications. A front load is also more aerodynamic than a rear load (see p. 35).

All these advantages make a front load a better choice than a rear load. To carry a front load well, the bike's geometry should be designed for the forces that the load imparts on the steering (see p. 85). Then a bike with a front load can handle as well as an unloaded bike.

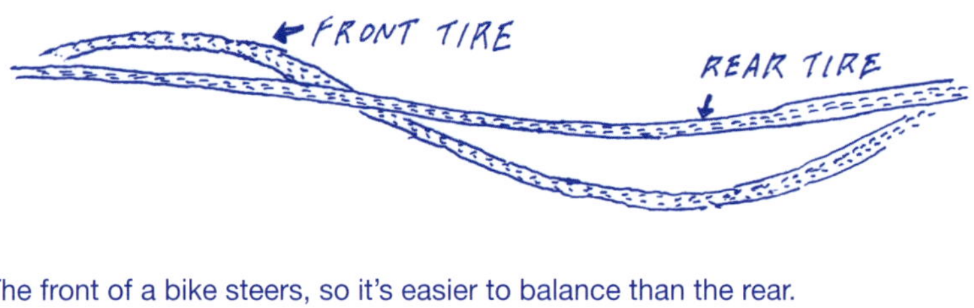

The front of a bike steers, so it's easier to balance than the rear.

Wheel Top vs. Low Rider

A handlebar bag offers convenience. It's easy to reach without dismounting, making it ideal for items that the rider needs to access during the ride. To minimize its effect on the bike's steering, the bag should be mounted low above the front tire and supported by a small rack. A porteur rack is useful to carry bulky loads over short distances, where the ease of loading outweighs the (small) negative effect on the handling.

For heavier loads and long rides, low-rider panniers are preferable. They mount on the fork close to the steering axis, which allows carrying even heavy loads with minimal effect on the bike's handling. The rack should be slightly offset from the steerer axis, so that its inertia stabilizes the steering. Otherwise, small movements of the panniers in relation to the rack translate into wobbles of the bike's steering.

For expedition-style touring, a rear rack can be used to carry light and bulky items that don't fit in the front panniers: sleeping bags, warm clothes and the like. A classic wheel-top rack allows placing the load further forward, but a rear low-rider lowers the bike's center of gravity for better cornering and easier rocking the bike when riding out of the saddle. It's hard to say which is better. Rear low-rider racks must be very stiff to minimize their negative impacts on the handling. This means that they should be custom-built for the bike and aren't available off the shelf.

Bikes are designed to be centered over their wheels, so panniers should be used in pairs, with the load equally distributed. A single-sided pannier makes the bike lean, which affects the steering and requires constant corrections.

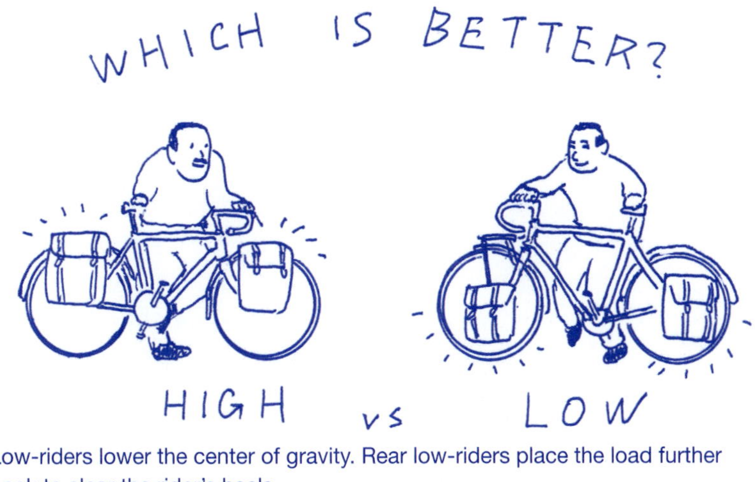

Low-riders lower the center of gravity. Rear low-riders place the load further back to clear the rider's heels.

Rack Design

Racks need to be stiff to avoid negative effects of the load on the steering. We guide our bikes with tiny movements of the handlebars. If our front load has a solid connection to the fork, the added inertia of the extra weight will just require a little more force when turning the handlebars – and a touring bike's geometry can (and should) be adjusted for that by increasing the wheel flop (see p. 103). If the rack flexes, then the load's inertia makes our bags stay in place at first when we turn the bars, only to spring back with a delay. This delay makes the handling unpredictable. For rear racks, flex increases the 'tail wagging the dog' effect already inherent in a rear load.

A good camping bike can be ridden no-hands even when loaded with 20 kg (44 lb) or more. Climbing out of the saddle will feel a bit different – less so with low-riders than with wheel-top racks – but it should be as intuitive and easy as on an unloaded bike. The best racks are made from steel tubing for stiffness. The rack's corners should be connected to the frame to prevent flex. Racks should be triangulated where possible – this means that no tube connects to the unsupported middle section of another tube. Sliders and adjustable brackets add weight and flex, especially if they are poorly designed.

If the rack slants outward at the bottom, gravity pulls the panniers toward the rack. This prevents wiggling and wobbling of the panniers. All these design elements combine to provide

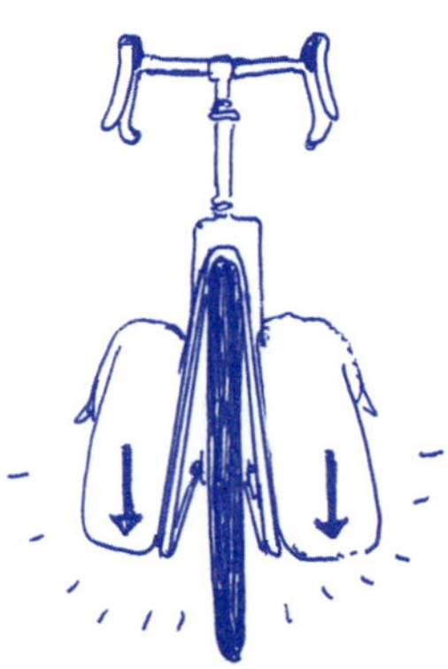

If the rack slants outward at the bottom, gravity pulls the panniers toward the rack and stabilizes the load.

a firm connection between load and bike, which is important to minimize the load's effect on the handling.

Racks also need to be strong: A broken rack can leave you stranded on a tour. A well-designed, well-made rack will last as long as the bike.

Bikepacking

The idea behind bikepacking is to fill various spaces within the outline of the frame with numerous small bags, rather than carry a few big bags on separate racks. Without bags sticking out to the sides, it's easier to ride through rough terrain or push the bike. Bikepacking is useful for touring on narrow singletrack: Distributing the load over multiple small bags keeps the handlebars light, which helps in tight corners, where the rider needs to turn the handlebars quickly. Perhaps the greatest advantage of bikepacking is that you can use any bike and don't need to mount racks on your fork and frame.

Bikepacking bags should be small and stay close to the outline of the bike to offer the best handling. This limits the capacity of most bikepacking setups. Large saddlebags 'wag the bike' when riding out of the saddle. For touring with significant loads on gravel or paved roads, a traditional pannier setup may be preferable.

Bikepacking allows carrying a camping load without racks.

Fenders

"Intermittent rain showers accompany my southward ride. There is no need to dress for the rain. Two layers of wool keep me warm, my fenders catch the road spray, and before I get really wet, the raincloud has moved on."

Nobody questions whether a Ferrari is a sports car, even though it has fenders. And yet fenders on a bike are often seen as a sign that it was not designed for performance. This needn't be so: If fenders are part of the original design – like on the Ferrari – rather than aftermarket add-ons, they won't detract from the performance.

You may not even plan to ride in the rain, but being able to ride on days with a 'chance of rain' in the forecast will greatly increase your options. And if it does rain, you will find that fenders with sufficient coverage keep road spray off your feet and drivetrain. Compared to the torrent of spray thrown up by your wheels, relatively little water falls from the sky. With good fenders, riding in the rain can be quite comfortable.

Mediocre fenders rattle and resonate on rough roads, and they offer only partial protection. Well-designed fenders simply disappear until it rains. Then they keep bike and rider dry and clean. The difference between excellent and mediocre fenders is almost as large as the difference between mediocre fenders and no fenders at all.

Fenders are especially important for gravel riding. Pavement dries within hours after the rain stops, but gravel can remain muddy for days. And even during a dry spell, fenders are useful because they keep dust off your bike.

Plastic vs. Aluminum vs. Steel

Plastic fenders – even carbon fiber ones – tend to be heavier than aluminum fenders, because they are more flexible and thus require stiffer, heavier stays. Stainless steel fenders are brittle and prone to cracking. Aluminum fenders not only are light, but also extremely durable when mounted correctly.

Effective fenders have an uninterrupted interior with rolled edges to keep the water inside. Many plastic fenders have mounting brackets riveted to their insides, where they

Whether sports car or all-road bike: Fenders improve versatility without decreasing performance.

form dams that divert water onto the fender stays, from where it drips onto the rider's feet. It doesn't sound like this should make a big difference, but if a drop of water falls on your feet every second, your shoes and socks get soaked in no time!

Coverage

The front fender should reach down to 12 cm (5") from the ground – just high enough that it does not get caught on curbs. A stiff rubber mudflap that extends beyond the fender down to 8 cm (3") above the ground keeps spray off your feet and your bike's drivetrain.

The forward end of the rear fender can extend a little below the bottom bracket to keep water off the chain in cross-winds. At the rear, a 'buddy flap' is appreciated by other riders during rainy group rides. So equipped, riding in the rain becomes not just bearable, but actually fun.

Ideally, the bike's frame is prepared for fender mounting:

- Fenders have 20 mm clearance to the tire.
- All bridges are equidistant from the wheel centers.
- Threaded holes face the tire to allow direct fender mounting.

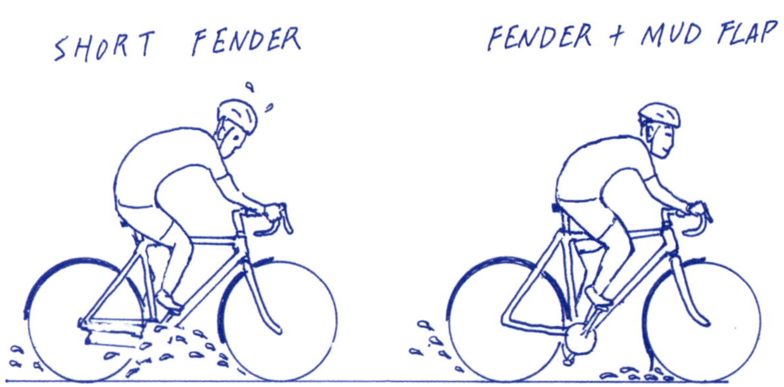

Front fenders with insufficient coverage spray water onto feet and drivetrain. A longer fender plus mudflap keeps feet and drivetrain clean and relatively dry.

Installing fenders properly prevents noise on rough roads and premature failure. Make sure that:

- Stays attach to the fenders with two bolts each to prevent flex.
- Front fenders have an attachment point in front of the fork.
- Fenders are shaped to follow the curve of the wheel before installation. Pulling fenders into shape with the stays results in inbuilt stresses and premature failure.

Fender Safety

When the front tire picks up rocks or small sticks, they hit the fender with great force. The fender can collapse, jam into the fork crown, and lock up the front wheel.

To prevent fender collapse, mount the fender far enough from the tire (20 mm) so that small stones and sticks can pass through without getting caught. Larger stones are too heavy to be picked up with much force, so they are less likely to collapse the fender. The fender itself should be as stiff as possible. Wider fenders that wrap around the tire more are stiffer. Fender collapse is not an issue with properly mounted aluminum fenders that are >50 mm wide.

As an alternative, a quick release for the fender stays can prevent the fender from jamming into the fork crown. This is a good idea for flexible plastic fenders and bikes with inadequate fender clearances.

Poorly mounted fenders have caused many serious accidents. It is better to get wet and muddy than take that risk!

Fender Summary

Well-designed and well-mounted fenders don't detract significantly from a bike's performance. They greatly increase the enjoyment of riding, not just when it rains, but also after rainshowers, when the road surface remains wet.

Lights

"Every thirty seconds or so, the road curves out of the beam of our headlights into a turn. We brake hard, turn in, adjust the radius as the headlight illuminates the road a bit further, then let the bike rise out of the lean for the next straight. We enjoy an exhilarating feeling of total focus, of the wind rushing by our heads, of living in the moment."

The sports car mentioned in the previous chapter also has lights, so you can drive it at night. Similarly, a fully-equipped bike allows you to venture out without fear of getting caught in the dark, if your adventure takes longer than planned. Or you may even enjoy rides in the mountains under a full moon, when the roads are free of traffic.

Generator vs. Battery

Today's generator lights have reached a level of maturity that makes them an excellent choice for spirited riding. Having lights at any time, with the flick of a switch, liberates you from worrying about charging batteries and how much run time remains before your lights turn dark. The resistance of the best generator hubs is too small to notice in most conditions. The added weight over a standard hub is less than that of a battery.

The best generator-powered headlights feature a carefully designed beam that provides an even light distribution for optimum visibility. The beam is cut off at the horizon, so oncoming traffic isn't blinded. When riding on multi-use trails, you should illuminate the bodies of pedestrians, but not their faces. And blinding car drivers isn't just impolite, but dangerous.

Beam Shape

The beam shape is far more important than the total light output. Our eyes adjust to the darkness, which is why even illumination is important. A bright spot right in front of the bike

Rally cars often race in the dark. Many performance bikes also have lights.

makes everything beyond difficult to see. An effective beam shape puts all the light on the road. This is more useful than a very powerful light that wastes most of its output by shining it into the sky (and into the eyes of oncoming traffic).

The best position for a front light is relatively low, so that potholes and road irregularities show up as shadows in the beam. The closer the light is to the rider's eye level, the harder it is to see bumps and holes. Mounting the light on the front axle is tempting at first, but the shadow cast by the front tire falls exactly where you need to see when rounding high-speed corners! A better position is on the fork crown or underneath a front rack (in a forward position so that the shadow of the front tire falls so far to the side that it isn't a problem).

Ideally, the angle of the headlight should be adjustable without tools on the road: Lower in the city to avoid blinding oncoming traffic, but a little higher during high-speed mountain descents, so that the light continues to illuminate your way even when you ride into a dip in the road. With the best generator-powered headlights, you can descend twisty mountain roads almost as fast at night as during the day.

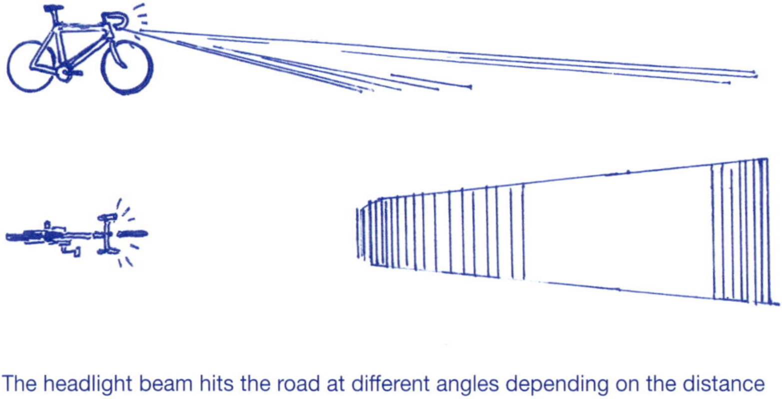

The headlight beam hits the road at different angles depending on the distance from the light. The beam should be shaped to even out the light distribution. At the top, the beam should be cut off to avoid blinding oncoming traffic.

Strobes and Helmet Lights

Flashing strobe lights draw attention, which can be useful during daytime. Flashing headlights are not intended for nighttime use, since they make it hard to see the road. And they really disturb drivers. The last thing you want is drivers coming toward you suddenly being unable to see where the road goes! They will err on the side of safety and steer toward the middle of the road – toward you!

Helmet lights are not a good idea as primary illumination. Especially in the city, riders move their heads constantly to scan the traffic that surrounds them. The light moves around so much that drivers might easily miss seeing it. Helmet lights can be useful to augment a bike-mounted light. They allow riders to look into corners when the bike leans and its headlight points toward the ground. They are also useful to see maps and cue sheets, as well as for repairs in the dark.

Shimmy

"My initial enjoyment of the bike was curtailed the first time I sat up and let go of the bars. The bike instantly shimmied wildly, and the resonance was accelerating rather than attenuating. It felt like the bike would shake itself to pieces!"

Shimmy or speed wobble is a rapid oscillation of the handlebars. When it occurs, it's usually while riding no-hands. There is no single factor that causes shimmy, which makes solving the problem difficult.

In almost every case, shimmy stops if the rider stops pedaling and rests one knee against the top tube. Few bikes shimmy while the rider holds the handlebars. For those that do, relaxing one's grip often stops the shimmy. Don't try to fight the shimmy by gripping the bars harder!

A cracked frame can cause shimmy. Frames rarely crack, but check this before you investigate anything else. Poor alignment of frame and fork can be another reason a frame shimmies. On the bike of a Bicycle Quarterly team rider, the shimmy disappeared after a new fork was installed.

Headset friction is a very important factor. A free-spinning headset with very little friction increases the likelihood of shimmy. The added resistance of a needle-bearing headset has greatly reduced the shimmy on several of our bikes.[54] The extra friction is too small to notice while riding, so the handling remains unchanged. As an added benefit, needle bearing headsets, with their larger bearing surface, last much longer than ball bearing headsets.

Handlebar bags increase the inertia of the front end of the bike, similar to a strong grip on the handlebars. The fixed points of the rider on the saddle and the handlebar bag are connected by the pivoting headset, which can lead to resonant frequencies and shimmy. Attaching the handlebar bag firmly to the handlebars greatly reduced shimmy on one BQ test bike.

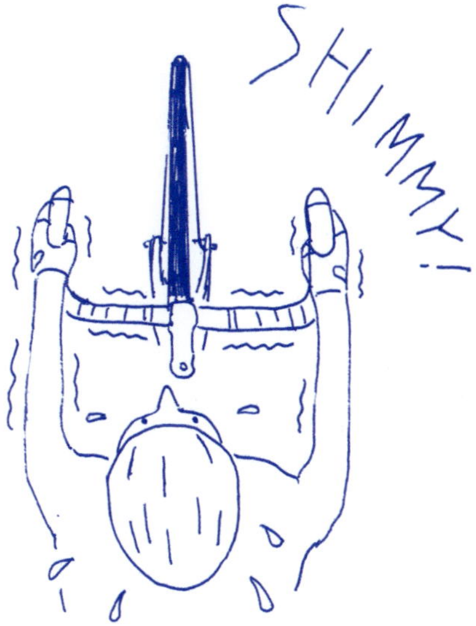

Shimmy can have many causes, making it difficult to resolve.

Wider tires and tires with more supple sidewalls appear to increase the likelihood of shimmy, because they allow the front wheel to oscillate with less damping than a narrower, stiffer tire. Extremely stiff frames may be less likely to shimmy, but flexible frames appear to be no more prone to shimmy than average ones. Top and down tubes with significantly different stiffness tend to prevent resonant frequencies from developing. For decades, most builders have used top tubes with smaller diameters and often thinner walls. It appears that this reduces the likelihood of shimmy. Conversely, bikes with oversize top tubes, but standard-diameter down tubes (making both tubes the same diameter) appear to shimmy more often than other bikes.[55]

Shimmy can be an intractable problem, because there are many variables, and they can interplay in many ways. Solutions that work on some bikes may not work on others, or even make things worse. Experimentation is the only way to resolve shimmy. If your bike shimmies even when you relax your grip on the handlebars, the first step is to replace the headset with a needle-bearing model. In most cases, this reduces shimmy to an acceptable level.

Unfortunately, many things that make a cyclotouring bike handle and perform well also increase its propensity for shimmy. If you want an absolutely shimmy-proof bike, you need a frame made from super-stiff tubing and run it with stiff, narrow tires. However, for most riders, such a bike would feel dead, and the harsh-riding tires would be slow and uncomfortable.

In our experience, well-made frames rarely shimmy excessively, even with low-trail geometries, flexible tubing, handlebar bags, and supple tires – as long as they are equipped with needle-bearing headsets. Sometimes, when riding no-hands at certain speeds, otherwise excellent bikes may develop a little wobble. Usually, pressing a knee against the top tube stops this, as does putting a hand on the handlebars. This is a small price to pay for the excellent all-around performance these bikes offer.

Cost

"When I bought my first custom bicycle, I was a student with little money. The builder, Hiroshi-san from Alps, told me not to buy expensive components, but to spend the money on train tickets and hostels. The bike was intended for travel, and you need money to travel."

It's nice to put together a cost-no-object dream bike – and the cost is far less outlandish than that of a dream car – but most cyclists have to work with a more realistic budget. This requires compromises. Should you buy middle-of-the-range parts everywhere? Or are there places where top-end components are really worth the money in extra performance and enjoyment? Where is the extra money well-spent, and what does it buy you?

The frame is one part where you don't want to compromise. It's the heart of the bike, and it determines how the bike rides and performs. It's also the part of the bike that is hardest to change. A good 'lively' frame makes your bike a joy to ride, whereas a budget frame will always feel sluggish.

Tires also make a large difference in the feel and performance of your bike. Fortunately, they are easy to change, and the wide tires of a good all-road bike spread their wear over a large rubber surface, so they last many thousands of miles. With wider tires, your tire budget can stretch to more expensive tires that offer a much better ride.

Factory-built wheels usually suffer from uneven spoke tension and poor stress relief, which will have spokes come loose or break prematurely. Hand-built wheels offer much better performance. If you buy a production bike, consider having the wheels completely rebuilt by an experienced wheelbuilder, using the stock parts. It may feel difficult to justify spending money on this, but your wheels will give you much longer and better service.

For the drivetrain, even relatively low-priced components from reputable makers offer good performance. For avid riders, the middle of the range offers the best value. Function and durability improve significantly with each step to a higher-level part. Once you reach mid-range components (Campagnolo Chorus, Shimano Ultegra), spending more buys you lighter weight, but it won't change the feel and performance of your bike. It's easy to justify compromises here.

With stems, handlebars and seatposts, reputable makers charge more, but their components also offer better safety. Beyond that, high-end parts save weight, and, in some cases, offer shapes that are more carefully studied for ergonomics and comfort. With fenders, racks and lights, the best parts usually feature superior design which results in better function and greater reliability.

My recommendation: Get the best frame you can afford and spend a little extra on hand-built wheels. Use mid-level components and choose safety-relevant parts from reputable makers. For fenders, racks and lights, it's worth spending the extra money on top-end parts, if you plan to carry luggage and/or ride in the rain and at night.

A great bike isn't cheap, but it's rare that someone regrets the extra money they've spent. It may require sacrifices in the short term, but it'll give you much joy in the long run.

Gravel and Cyclocross

"The corners on this descent require only a touch of brakes before we commit, balancing our bikes at the edge of grip. Once in a while, the front tire slips, but the slides are easy to catch. This ability to go to the limits of grip and beyond is why riding on gravel is so exhilarating."

Gravel riding is popular, and for good reason. Many gravel roads are almost free of traffic. The scenery tends to be nice. And your bike feels different on gravel. The tires slide a little. It's a bit like skiing, where your skis float on the snow. On gravel, the boundary between grip and slide is fluid, and that sense of freedom can be exhilarating. It's also a great way to learn the feedback from your bike: Slides happen sooner, but they are less likely to result in a fall. Once you know what it feels like when your bike is about to slide, you'll feel much safer on pavement, too.

Riding on gravel does not require a special bike – we used to 'underbike' on gravel roads and forest tracks on 28 mm tires. With the right equipment, gravel is faster and more fun.

All-Road vs. Monstercross

There is much confusion about what a gravel bike is, because gravel riding defies easy categorization. Most gravel bikes fall into two camps: all-road bikes that are essentially road bikes with wide tires, and monstercross bikes with geometries inspired by mountain bikes. It's not that one is better than the other – which is right for you depends on how you plan to ride. Do you see yourself flying across gravel roads, balancing the tires at the limits of grip during fast descents? Or do you see the open roads only as connectors between sections of technical singletrack, where the challenge is to 'clear' obstacles and carry speed through tight and twisty trails?

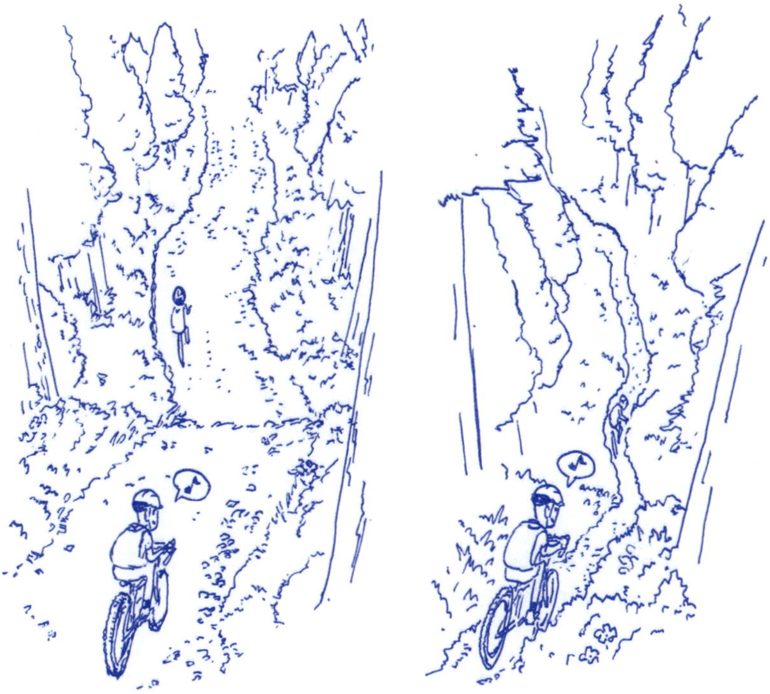

All-road bikes are for fast riding on gravel roads with sweeping turns (left). Monstercross bikes excel in technical terrain with tight turns (right).

If you ride fast on gravel roads, you'll want an all-road bike – a road bike with wider tires. Because gravel roads are roads, after all. If you want a gravel bike for technical terrain and singletrack, you'll need a monstercross bike with a mountain bike-inspired geometry. The difference boils down to how much you turn the handlebars.

At high speeds, you flow through corners. Your handlebars turn very little as you set up the bike for a corner by countersteering, then steer into the lean, before righting the bike for the next straight. A road bike position – relatively low and with significant weight on the front wheel – works well, and you'll want a frame with a relatively steep head angle and relatively little trail (and little wheel flop).

In technical terrain, you need to turn sharply at low speeds. You'll rotate your handlebars quite far, and so you need more 'power steering' – a shallower head angle, more trail and hence more wheel flop. You'll also want a more upright riding position with less weight on the front wheel. As we've discussed in the chapter on front-end geometry (p. 107), an 'in-between' mid-trail geometry does not work well, because it has too much trail for quick changes in direction and not enough wheel flop to turn the bars and overcome that resistance.

Gravel Tires

Road tires work well on gravel, if they are wide enough, because tire tread makes little difference: What limits your grip isn't the tire sliding on the topmost layer of stones, but the stones sliding on each other. Without a firm surface to dig into, knobs don't really improve grip on most gravel surfaces.

Punctures are relatively rare on gravel, because debris gets pushed into the soft road surface rather than into the tire. However, sharp rocks can cut the sidewalls of high-performance tires, and many gravel tires have reinforced casings. That also makes them slower and less comfortable, so it may be worth experimenting with high-performance tires. Let the bike move around under you, so the tires can bounce off the rocks rather than being pushed into them, and you may find that supple high-performance tires are durable enough for gravel riding.

Cyclocross

Cyclocross bikes are similar to gravel bikes – both are road bikes with wide tires. For decades, racers used their normal road bikes, equipped with knobby tires, for cyclocross. Purpose-built cyclocross bikes became popular when the tires of road bikes became narrower and frame clearances were too tight to fit wide tires.

Cyclocross races tend to be muddy, which requires a few changes from gravel bikes. The most obvious one is knobby tires. Cyclocross includes lifting your bike across barriers and even carrying your bike up hills that are too steep or muddy to ride. A horizontal top tube makes the bike easier to shoulder and carry during the dismounts that make 'cross so much fun.

There are many choices among gravel and 'cross bikes. Whereas most road bikes tend to be similar, the differences between gravel bikes really change how they ride and perform. Make sure you get the right bike for your style of riding!

Tandems

"Riding a tandem down a long mountain-pass descent is almost like cheating. We are tucked and flying down the moderate grades. We easily spin out a 54-13 gear on the sections where the gradient decreases to relatively mild levels. The altitude gains that took big chunks of time spin away in mere minutes."

Tandems used to be popular because they offer a great way to share the joys of cycling as a team. Key to good teamwork is a similar pedal stroke of both riders. With the right team, tandems fly on the flats – imagine a cyclist with twice the power, but almost the same wind and rolling resistance. They also climb as fast as single bikes – two riders may weigh twice as much, but they also have twice as much power. And during descents, the extra weight – without additional wind resistance – becomes a big advantage.

Tandems set the fastest times in the first three post-war Paris-Brest-Paris randonneur brevets over a very hilly, 1200 km-long course. Flying along on a tandem, pedaling in perfect unison with a partner, is a truly memorable experience.

If the riders' pedal strokes don't match, much energy is lost in the synchronization of the two riders. This is especially noticeable uphill, where the aero advantage of the tandem vanishes. If your focus is on spending time with your tandem partner rather than performance, then this matters little. Just pick a flattish course and enjoy the ride!

Back to Basics

What you'll read in this chapter runs counter to what many makers and riders of current tandems will tell you, yet it's based on riding many thousands of miles on more than a dozen tandems, both on the front and in the back, with ten partners. The best classic tandems from French makers like Rene Herse or British ones like Jack Taylor were remarkably refined machines.[56] They were easy to ride, comfortable and fast.

Tandems did not adapt well to the narrower tires that became popular during the latter half of the 20th century. With the rear rider (stoker) unable to see the road ahead, comfort is doubly important, and even 42 mm-wide tires are marginal on a tandem. Using suspension seatposts to regain some of the comfort lost to narrow tires resulted in the stoker bobbing up and down and getting out of sync with the captain.

Classic low-trail geometries for tandems rely on the pneumatic trail of wide tires for stability. Other geometries, often derived from single bikes, resulted in too much wheel flop and made many tandems unstable when the stoker moved, for example to reach for a water bottle. Misconceptions about tandem sizing resulted in tandems that were longer than necessary, making them more flexible, more difficult to maneuver and less aerodynamic.

Returning to the old designs allows returning to machines that are faster, more comfortable, easier to handle – and more beautiful. However, it is important to remember that tandems, even more than single bikes, are complex machines. Many old 'budget' tandems had flexible frames made from tubing for single bikes, poor-handling single-bike geometries, and budget components not up to the loads imparted by a two-person team. These tandems were truly awful, and they made it easy to overlook the sophisticated – and rare – machines from the best builders.

Tandem Design

We've already covered the specifics of tandem frame geometry on p. 105, so we'll just add a few details here. Tandems are so long and carry so much weight and power that even oversize tubes flex significantly. Thus, frame stiffness is more important on a tandem than on a single bike. And if stiff chainstays give single bikes better performance (see p. 65), what does this mean for tandems, where the distance from the front bottom bracket to the rear wheel measures over 100 cm (40")?

The best way to stiffen a tandem is to shorten the frame. Fortunately, tandem frames can be relatively short, because the rear rider does not need to crouch in the wind, so they can adopt a more upright position. An added benefit is that they can see over the shoulder of the front rider.

Frame Sizes

When sizing the rear compartment of a tandem, the relative sizes of captain and stoker must be considered. If the stoker is of smaller build than the captain, their head will be uncomfortably close to the captain's bottom, so the rear top tube needs to be longer to create extra room. If both riders are of similar height, the stoker's head will be above the captain's back, and the rear top tube can be relatively short.

Jack Taylor built more than a thousand custom tandems, and nearly all have rear top tubes that are 60 cm long. This may come as a surprise, but for most teams, a rear top tube length of about 60 cm works well, almost regardless of the rear rider's height. A taller stoker's head is above the captain's back, so they can ride comfortably with a relatively short reach. A shorter stoker's head is lower, and thus needs a little extra room to the captain's bottom. But their torso is shorter, too, so the resulting top tube length comes out to a similar value.

If both captain and stoker are short, the rear top tube can be shorter than the 'standard' 60 cm.

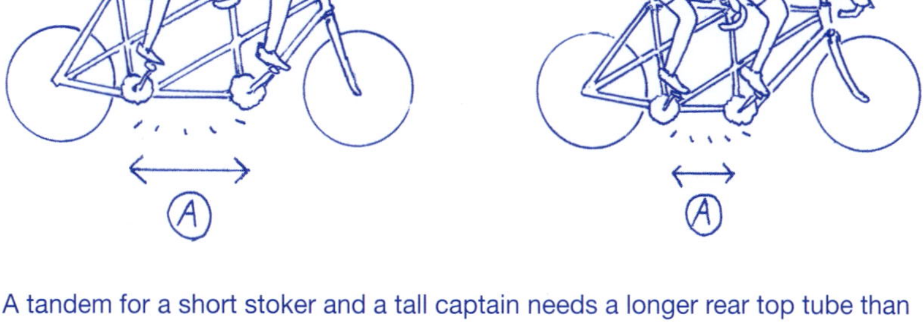

A tandem for a short stoker and a tall captain needs a longer rear top tube than a tandem for two riders of similar height.

Frame Stiffness

A short rear compartment will help stiffen the frame. It also improves the aerodynamics of the tandem and hence its speed, because both riders form a single body that moves through the air. To shorten the chainstays further, some tandem builders have curved the rear seat tube, which also shortens the wheelbase slightly for better maneuverability. Having ridden two René Herse Chanteloup tandems with curved rear seat tubes, I can attest to their excellent performance and handling.

To increase the stiffness of tandem frames, additional tubes can be added. Double triangulation appears to be the most effective way to stiffen a tandem frame. The bracing tubes run from the center of the frame downward and outward, so that they turn the two-dimensional frame into a three-dimensional structure. Having ridden otherwise similar tandems with and without the double triangulation, it is easy to notice the additional rigidity that comes from the four long diagonal stays. The result is more surefooted handling and better performance. Adding tubes in the same plane as the frame's main tubes will make the tandem stiffer vertically, but it's less effective in improving performance.

Triangulation is the best way to add much-needed stiffness to a tandem frame.

Most tandems today use short-chain transmissions with the drive to the rear wheel off the rear cranks. A long-chain transmission, with the drive coming off the front cranks, eliminates the large rear chainrings. This allows the chainstays to be shorter and still fit around wide tires. Disadvantages of the long-chain transmission are a longer (and thus heavier) chain and the need for a rear derailleur with a strong spring to provide enough chain tension.

Tandem Summary

Tandems have lost popularity because they are difficult to make. The difficulty of finding a partner whose pedal strokes harmonize means that they sell only in small numbers. Without mass producers promoting them, the mainstream cycling media ignore them. And yet tandems offer a unique teamwork aspect that can add greatly to the enjoyment of cycling. And then there is the incredible speed that a good tandem can maintain...

Tandem design is complex, and simply lengthening a single bike and beefing up its frame will not yield optimum results. Tandem geometries must consider the extra weight causing more wheel flop, and tandem frames must balance the fit for two riders with the need to keep the frame as short and stiff as possible. Building a great tandem is far more difficult than building a great single bike, but the rewards are worth the extra effort.

Choosing a Custom Builder

"It's not easy to make a good bicycle. A great bike is sophisticated. The magic is invisible, but you feel it when you ride."

There are many excellent production bikes. With super-light carbon fiber frames requiring expensive molds, it can be argued that the best production bikes offer performance – or at least light weight – that cannot be obtained from a custom bike. And yet there are many custom builders, and there are good reasons to order a custom bike.

For some riders, it's simply a matter of fit. If your body's proportions fall outside the 90% for which production bikes are sized, you are better off on a custom frame, rather than trying to adapt an ill-fitting production bike with an extra-long or -short stem and seatpost.

For other riders, it's about personalization. Getting something that is unique and made just for you has an appeal of its own, and the difference in cost is often remarkably small, since small builders have little overhead and no advertising budget that gets rolled into the price of the bike.

There is another reason why many of us ride custom bikes: Few production bicycles are available with integrated fenders, rack and lights. For the ultimate real-world bicycle, you are almost forced to order a custom bike.

Working with a true artisan is fun, and bicycles are one of the few areas where custom-crafted, hand-made objects remain relatively affordable. For the price of a decent used car, you can get an amazing bicycle.

Getting the Perfect Custom Bike

When you order a custom bike, the most important thing is choosing the right builder, so that you really get the bike of your dreams. Buying a custom bike is an act of faith: You cannot test-ride your new bicycle.

Most established builders are skilled craftspeople. Yet disappointments occur, usually because customers have selected the wrong builder for their bike. Instead of trying to find a builder you like, choose the type of bike you want. Is it a racing bike, a fully-equipped randonneur bike, a city bike with an upright position, or a tandem with camping racks? Think of the level of detail you desire. Should racks, fenders and lighting be integrated into the bike? Then you need a true constructeur, who designs the entire bike as an integrated unit. Or do you prefer accessories that you can add and remove as you like? Then a framebuilder will be fine, who will add braze-ons and then let you figure out which accessories you can mount and how.

To choose the right builder, look at their show bikes and personal machines to see where their heart lies. Don't order an all-road bike from a builder who rides a fixie, or a 650B randonneur bike from a builder who believes in narrow tires.

If the builder is familiar with the type of bike you are ordering, he or she is more likely to get all the crucial details right. Especially on a bike with integrated racks, fenders and lights, there are a lot of details that make the difference between a superb machine that provides decades of trouble-free performance, and one with small problems that will bother you on almost every ride.

Look for experience – not only in building bikes, but also in designing them. If your bike is the first one of its type, it will be a prototype, not a final product. You want a builder who has plenty of experience building bikes exactly like the one you are ordering. The best bikes have evolved based on feedback from avid riders. The ideal builder works closely with riders like you, and with each bike, learns more about what works and what could be improved. In the old days, many builders sponsored teams. This was more than just advertising. It allowed builders to improve their bikes through feedback from experienced riders.

Make sure your builder shares your ideas. If the builder suggests a certain geometry or component choice, accept it or choose a different builder. Do not try to convince your builder to make a bike they do not believe is right for you.

A good builder will ask questions about your preferences and riding style, so they can craft the perfect bike for you. Once you have chosen a builder, you have to trust them. Don't try to micromanage all the details, but make sure you communicate your wishes clearly. Each builder has their own style, and most will ask you for your input as needed. Many builders want to meet their customers and see them ride. It's a great way to ensure the builder understands your riding style and preferences, and it's a great experience for you. Even if it is a long trip, it will be a lasting memory.

When your dream bike arrives, it is likely to be the best bike you ever have ridden. Enjoy it, tune in to its personality, and learn how to get the most out of it. A great custom bike may be expensive, but if it's the right bike for you, it will give you joy for decades.

Riding a Performance Bike

"A good bicycle will tell you how to descend. Just follow the bike. Look in the direction you want to go. Feel the gravity. Feel the road through the bicycle. Relax."

This book discusses how to make a bike that performs well, and how to choose a bike that is right for you. We haven't talked much about the most important component: you, the rider. Bicycles are unlike other vehicles: They are lighter than their riders, and their riders are also their engines. Bicycles are fundamentally different from vehicles like cars or motorbikes. A better comparison is a mechanical body part – it must work in perfect harmony with the person. That is why the flex characteristics of the frame should match the rider's pedal stroke, and vice versa: To get the most from a bike, the rider must tune his or her pedal stroke to the bike.

A bicycle's true performance potential can be realized only when rider and bike are working in unison. "Listen to the bike!" is probably the best advice about riding technique I've ever heard. Guide the bike with a light touch on the handlebars. A good bike doesn't need to be forced (except perhaps in very strong crosswinds), but it will react to very small inputs. Remember that bikes are self-stable – they don't change course unless you do something: pull on the handlebars, shift your weight, etc. Relax, and the bike will go where you guide it. Work on pedaling smoothly without bopping up and down on the saddle. Try to ride in a straight line by riding on the white 'fog line' at the edge of the road. You will find that this works best when you look far ahead and, Zen-like, don't focus on the line at all.

Look where you want to go, and the bike will follow. This means: Don't focus on potholes, but on the path between them. When cornering, turn your head in the direction of the turn, and the bike will round the corner smoothly and without jerkiness. Trust the bike, and it'll work with you.

Over time, you'll become one with your bike. You'll move with the grace of a ballet dancer, and you'll become the heart of a finely tuned machine. Improving your skills is a life-long process, and it will make your cycling more enjoyable – and also faster and safer. This process will keep your cycling fresh and exciting for many years to come.

Thank you

I am greatly indebted to the many craftsmen who have shared with me what they know about making bicycles. Some of them I met only through studying their bikes, like René Herse, Jo Routens, Cino Cinelli and André Reiss (Reyhand). I was fortunate to spend significant time in the workshops of others, observing them or working with them. Glenn Erickson, Bill Davidson, Ernest and Olivier Csuka (Cycles Alex Singer), Jean Desbois (Cycles Rene Herse), H. Hirose, and the craftsmen at Toei, Mr. Yamada and Mr. Kobayashi, have taught me much of what I know about bicycles.

Peter Weigle, Jeff Lyon and Matthew Grimm (Kogswell) contributed to our research by shaving the tread off prototype tires (Peter) and building a number of frames (Jeff, Matthew) and forks (Peter). Without this generous help, testing our hypotheses about tire performance, steering geometry and frame stiffness would have been difficult.

Jim Papadopoulos, Frank Berto and Raymond Henry have freely shared their insights. I owe great thanks to the Bicycle Quarterly Team – a group of avid cyclists without whom this research would not have been possible. Mark Vande Kamp has provided his expertise both on the bike (he's an amazing rider) and off (he has a Ph.D. in mathematics). Lael Wilcox and Ted King have ridden with me to the end of the road and back, and often our conversation has been about what we want in a bike.

Special thanks goes to my editor, Natsuko Hirose, a keen cyclist herself. Her vision turned a collection of technical data into a coherent and beautiful book, and her insights into all things cycling provided valuable perspectives. Seeing her descend twisty mountain roads reminded me that all technical research has only one purpose: to make cycling even more fun.

Notes

Part 2

Aerodynamic Resistance

1 Hale, J., J. Heine, M. Vande Kamp and A. Wetmore, 2007: Aerodynamics of Real-World Bicycles. Bicycle Quarterly 21 (Vol. 6, No. 1), p. 1.
2 Lorimor, E., J. Hale, J. Heine, A. Wetmore and M. Mark Vande Kamp, 2008: The Aerodynamics of Drafting and of Different Bike/Rider Types. Bicycle Quarterly 23 (Vol. 6, No. 3), p. 45.

Rolling Resistance

3 The Performance of Tires. Bicycle Quarterly 17 (2006), p. 1.
4 Tire Pressure and Performance. Bicycle Quarterly 43 (2013), p. 26.
5 Pradko, F., and R. A. Lee, 1966: Vibration Comfort Criteria. Society of Automotive Engineers Paper No. 660139.
6 Comfort Equals Speed – Quantifying Suspension Losses. Bicycle Quarterly 29 (2009), p. 6.
7 Minimizing Suspension Losses – Equipment that Makes You Faster and More Comfortable. Bicycle Quarterly 29 (2009), p. 8.
8 Tire Performance 101. Bicycle Quarterly 43 (2013), p. 34.
9 Do Larger Wheels Roll Faster? Bicycle Quarterly 36 (2011), p. 10.
10 see 6 (Comfort Equals Speed, Bicycle Quarterly 29, 2009).

Weight

11 Bicycle Acceleration Model developed by Forbes Bagatelle-Black. Rider: 60kg; bikes: 6kg; A: 0.5m^2; c_d: 0.92; Crr: 0.005; rims/nipples: Bike 1: 722g; Bike 2: 980g; tires/tubes: Bike 1: 560 g; Bike 2: 870 g.

Bearing Resistance

12 Zinn, L., 2007. VeloNews Vol. 35, No. 21, p. 40.
13 Spicer, J. B., C. J. K. Richardson, M. J. Ehrlich, J. R. Bernstein, M. Fukuda and M. Terada, 2000: Efficiency of chain drives. Human Power 50, p. 3.

Frame Flex

14 Shop Visit: Cherubim. Bicycle Quarterly 69 (2019), p. 88.
15 Frame Stiffness Matters. Double-Blind Test Confirms that Planing is Real. Bicycle Quarterly 24 (2008), p. 42.
16 Planing and Oversize Tubing. Bicycle Quarterly 28 (2009), p. 88.
17 Frame Stiffness and Pedal Stroke. Bicycle Quarterly 28 (2009), p. 88.
18 Heine, J., 2008: The Competition Bicycle: A Photographic History. Bicycle Quarterly Press, Seattle, 176 p.
19 Frame Stiffness: The Balance of a Frame. Bicycle Quarterly 35 (2011), p. 50.

20 Test: Trek 2100C – Suspension for Comfort and Speed. Bicycle Quarterly 16 (2006), p. 38
21 For example: Bike Test: Jeff Jones 29er. Bicycle Quarterly 51 (2015), p. 14.

Biomechanics

22 Disley, B. X. E. St. J., 2013: Q Factor in Cycling: Kinematic and Physiological Effect. University of Birmingham, UK. See also: Disley, B. X., 2018. Q Factor, Performance and Injury. Bicycle Quarterly 66, p. 58.

Handling

23 Balancing and Cornering on a Bike. Bicycle Quarterly 34 (2010), p. 44.
24 Papadopoulos, J., 2004. Steering and Balancing. In: Wilson, D. G., 2004: Bicycling Science. MIT Press, Cambridge, 477 p.
25 Front End Geometry for Different Speeds, Loads and Tire Sizes. Bicycle Quarterly 11 (2005), p. 28.
26 How to Design a Well-Handling Bicycle. Bicycle Quarterly 19 (2007), p. 42.
27 Optimizing Bicycles for Carrying Heavy Loads. BQ 19 (2007), p. 37.
28 Wheel Size and Bicycle Handling. Bicycle Quarterly 31 (2010), p. 14.
29 Gyroscopic Stability. Bicycle Quarterly 34 (2010), p. 69.
30 Foale, T., 2006: Motorcycle Handling and Chassis Design – The Art and the Science. 2nd Ed.

Comfort

31 Moulton, A., 2009. From Bristol to Bradford-on-Avon – a lifetime in engineering. The Rolls-Royce Heritage Trust, 320 p.

Part 3

Tires

32 see 3 (Bicycle Quarterly 17, 2006); 8 (Bicycle Quarterly 43, 2013); Testing the Performance of Tires, Part 2. Bicycle Quarterly 19 (2007), p. 18; Tire Performance on Smooth Roads. Bicycle Quarterly 43 (2013), p. 23.
33 What Makes a Tire Fast? Bicycle Quarterly 47 (2014), p. 54.
34 see 3 (Bicycle Quarterly 17, 2006).
35 Are Wider Tires Slower on Smooth Roads? Bicycle Quarterly 58 (2016), p. 70.
36 see 3 (Bicycle Quarterly 17, 2006).
37, 38 see 35 (Bicycle Quarterly 58, 2016).
39 Clinchers vs. Tubulars – Which is Faster? Bicycle Quarterly 43 (2013), p. 30.
40 Latex Tubes. Bicycle Quarterly 74 (2020), p. 102.
41, 42 Testing Tires: How Fast Do They Roll. Bicycle Quarterly 73 (2020), p. 70.

43, 44 see 3 (Bicycle Quarterly 17, 2006).
45 Aaron, M., 1988. Importance of Real-World Results. Bike Tech, October 1988, p. 5.
46 see 3 (Bicycle Quarterly 17, 2006).

Wheels

47 Brandt, J., 1993. The Bicycle Wheel, 3rd Edition.
48 Test: Toei Randonneur. Bicycle Quarterly 28 (2009), p. 56.

Fork

49 Measuring the Flex of Fork Blades. Bicycle Quarterly 23 (2008), p. 24.
50 see 7 (Bicycle Quarterly 29, 2009).

Brakes

51 Testing Brakes. BQ 37 (2011), p. 28.

Drivetrain

52 Berto, F., 2001: Hub Gear vs. Derailleur Gear Efficiency. Cycle History 12.

Racks and Luggage

53 see 22 (Papadopoulos, 2004).

Shimmy

54 Curing Shimmy on a Bike. Bicycle Quarterly 23 (2008), p. 44.
55 Top Tubes, Planing and Shimmy. Bicycle Quarterly 23 (2008), p. 43.

Tandems

56 The Fascinating Tandems of France. Bicycle Quarterly 7 (2004), p. 1.

Index

Jan Heine became fascinated with bicycles when he started racing in college. By the time he completed his Ph.D., he was racing alongside professionals preparing the Tour de France. Deciding that a pro racing career was not for him, Jan's interests turned to long-distance adventures, and he was the fastest North American rider in the epic 1200 km Paris-Brest-Paris in 2007. Around the same time, he started exploring the long-forgotten gravel passes of the Cascade Mountains and realized that the then-current bikes weren't well-suited for spirited gravel adventures. The logical next step was to figure out how to make them better. Over the last two decades, Jan has combined his scientific background and his real-world riding experience to develop tires and components for Rene Herse Cycles.

Published by Bicycle Quarterly Press
Seattle, WA, USA
www.bikequarterly.com

Editor: Natsuko Hirose
Graphic Design: There There
Illustrations: MIYOSHI

Printed and bound in China

Library of Congress Cataloging-in-Publication Data
Names: Heine, Jan, author | King, Ted, foreword
Title: The All-Road Bike Revolution / Jan Heine
Description: First U.S. edition, second printing | Seattle, Washington: Bicycle Quarterly Press, 2021 | Includes bibliographic references and index
Identifiers: LCCN 2020945200 | ISBN 9780976546054 (hardcover)
Subjects: Bicycles and bicycling | Technology | Sports